Challenging Future Practice Possibilities

Practice Futures

VOLUME 1

Series Editor

Joy Higgs (*Education, Practice and Employability Network, Australia*)

Advisory Editorial Board

Steven Cork (*Australian National University, Australia*)
Geoffrey Crisp (*University of New South Wales, Australia*)
Debbie Horsfall (*Western Sydney University, Australia*)
Will Letts (*Charles Sturt University, Australia*)

Scope

The way people act and work – and are enhanced or replaced by technology – in employment and practice settings in the future, will inevitably evolve incrementally or radically. This series considers *probable, possible* and *preferable* practice futures and employability, along with accompanying educational influences and support. Wisdom is a key dimension of our discussions of practice and education. The books in this series examine directions that are currently underway, future visioned and at the edge of imagination in transforming practices. The authors reflect on how these transformations are being or could be influenced by many factors including changes in society, nations and global connections, in the physical environment, workplaces (physical and virtual) and in the socio-economic-political contexts of the world and its nations. The authors in this series bring a rich range of practical and academic knowledge and experience to examine these issues. Through the conversations in these books readers can enter into these debates from multiple perspectives – work, organisations, education, professional practice, employability, society, globalisation, humanity, spirituality and the environment.

The titles published in this series are listed at *brill.com/pfp*

Challenging Future Practice Possibilities

Edited by

Joy Higgs, Steven Cork and Debbie Horsfall

BRILL
SENSE

LEIDEN | BOSTON

All chapters in this book have undergone peer review.

The Library of Congress Cataloging-in-Publication Data is available online at http://catalog.loc.gov

ISSN 2665-9263
ISBN 978-90-04-40077-1 (paperback)
ISBN 978-90-04-40078-8 (hardback)
ISBN 978-90-04-40079-5 (e-book)

Copyright 2019 by Koninklijke Brill NV, Leiden, The Netherlands.
Koninklijke Brill NV incorporates the imprints Brill, Brill Hes & De Graaf, Brill Nijhoff, Brill Rodopi, Brill Sense, Hotei Publishing, mentis Verlag, Verlag Ferdinand Schöningh and Wilhelm Fink Verlag.
All rights reserved. No part of this publication may be reproduced, translated, stored in a retrieval system, or transmitted in any form or by any means, electronic, mechanical, photocopying, recording or otherwise, without prior written permission from the publisher.
Authorization to photocopy items for internal or personal use is granted by Koninklijke Brill NV provided that the appropriate fees are paid directly to The Copyright Clearance Center, 222 Rosewood Drive, Suite 910, Danvers, MA 01923, USA. Fees are subject to change.

This book is printed on acid-free paper and produced in a sustainable manner.

CONTENTS

Preface   vii

Acknowledgements   ix

Part 1: Grappling with Practice Futures

1. Exploring Practice in Context   3
   Joy Higgs

2. Thinking the Unthinkable: Challenges of Imagining and Engaging with Unimaginable Practice Futures   17
   Steven Cork and Debbie Horsfall

3. Plausible Practice Futures   29
   Steven Cork and Kristin Alford

4. The Impact on Practice of Wicked Problems and Unpredictable Futures   41
   Peter Goodyear and Lina Markauskaite

5. The Changing Face of Work: Considering Business Models and the Employment Market   53
   Paul Whybrow and Asheley Jones

Part 2: Practice and the Common Good

6. Re-claiming Social Purpose and Adding Values to the World around Us   65
   Debbie Horsfall and Joy Higgs

7. Our Place in Society and the Environment: Opportunities and Responsibilities for Professional Practice Futures   79
   Steven Cork

8. Practice Futures for Indigenous Agency: Our Gaps, Our Leaps   91
   Sandy O'Sullivan

9. Changing Work Realities: Creating Socially and Environmentally Responsible Workplaces   101
   Rosemary Leonard and Margot Cairnes

10. Towards Future Practice in Socio-political Contexts   113
    Megan Conway and Joy Higgs

Part 3: Pursuing Practice Futures

11. The Place of Agency and Related Capacities in Future Practices   129
    Franziska Trede and Joy Higgs

CONTENTS

12. Employability and Career Development Learning through Social Media: Exploring the Potential of LinkedIn 143
    Ruth Bridgstock

13. Re-imagining Practice Structures and Pathways: Starting to Realise Tomorrow's Practices Today 153
    Joy Higgs and Daniel Radovich

14. Freelancing, Entrepreneurship and Inherent Career Risk: An Exploration in the Creative Industries 167
    Noel Maloney

15. Young People's Hopes and Fears for the Future 177
    Steven Cork and Jennifer Malbon

16. Facing Recruitment Challenges: Entering Workplace Practices 187
    James Cloutman and Graham Jenkins

17. PhDs and Future Practice 199
    Bernadine Van Gramberg

18. Educational Innovations: Preparing for Future Work 209
    Asheley Jones

19. Otherness in Practice (in the Health Professions) 219
    Janice Orrell and Julie Ash

20. Workplace Innovations and Practice Futures 229
    Thomas Carey, Farhad Dastur and Iryna Karaush

Part 4: Reflections

21. Reflections about Work: What Might Be My Future Practice Roles? 245
    Joy Higgs

Notes on Contributors 253

PREFACE

Change is a certainty and the pace of change in the 21st century is omnipresent with no part of our lives remaining unaffected. This includes changes in technology, work, political systems, how goods and services are provided, how we are cared for and care for others, how our children grow, the nature of our relationships and how we communicate with each other, what we eat and how that food is produced, changes in the natural world, our notions of social justice and fairness, how we warm and heat our houses, public buildings and shared spaces, even changes to the very air that we breathe. There are many reactions to this fast pace of change: excitement, curiosity, greed, exploitation, hope, anxiety, alienation, blame, avoidance, and bewilderment. And, we hope that our responses include a sense of possibility that yes the world could be a better place for future generations. What is certain is that it will be a vastly different place that the next generations including our children and grandchildren inhabit and it will be vastly different in terms of what, where and how we work. While the rate of change may possibly be overwhelming we do not want to be passive or immobilised bystanders, uncritical followers or colluders in changes which are not for the greater good. (Practice Futures Forum)[1,2]

The title of this book is *Challenging Future Practice Possibilities*. We chose the term *practice* to refer generally to what people do but more specifically to what they do in work and professional practice/work. Future practice is examined through its emerging trends, possibilities and challenges. Some of these matters are occurring around us while others are being born in our imaginings. Still others are unimaginable today. Within this space we have invited authors to explore these issues across a wide range of settings and interests under these headings: Grappling with practice futures, Practice and the common good, Pursuing practice futures and Reflections.

On a personal note, we invite you to reflect on your future practice/work and what difference you will seek to make to your own role and practices as a practitioner, worker, manager and client in the future worlds of practice, work, organisations, commerce, service, industries, communities and societies. How can your practice improve? What differences will you make to your workplaces, practice communities, clients and society? What influences will you create and respond to in these endeavours? How will you help shape the prospects of future generations?

Joy Higgs, Steven Cork and Debbie Horsfall

NOTES

[1] https://www.practicefutures.com.au/
[2] https://www.epen.edu.au/news

ACKNOWLEDGEMENTS

The editors wish to acknowledge the excellent support received by Kim Woodland for her invaluable book management and copy editing, and Jennifer Pace-Feraud for her role in quality checking.

PART 1

GRAPPLING WITH PRACTICE FUTURES

JOY HIGGS

1. EXPLORING PRACTICE IN CONTEXT

This book is about practice futures and much of it deals with future visions and challenges. But what of practice itself? What is it and why does it matter to us? First, in order to contemplate a future, we need to ask: the future of what? Scholars, looking at the topic abstractly, need to understand how the authors are portraying this complex phenomenon. Workers across the professions plus other industries and occupations will be interested in the practical, personal and economic aspects of their jobs, their businesses and their organisations. Leaders in society, business and organisations are invited to reflect in this book on triple-bottom-line matters and the impact of their decisions on future generations and society.

The phenomenon of practice is a social construct in that to understand it we need to give meaning to it: this is a social rather than individual process of meaning making and attribution. A (collective) practice comprises ritual, social interactions, language, discourse, thinking and decision making, technical skills, identity, knowledge, and practice wisdom, framed and contested by interests, practice philosophy, regulations, practice cultures, ethical standards, codes of conduct and societal expectations. For instance, if we talk about another construct, 'the *practice* of democracy' we could examine its dimensions and enactment through the living words of language and the documented words of the discourse of democracy in general, or the various models through which it is realised or enacted in different nations. If we talk about 'the *practice* of medicine' we could define it broadly in comparison to the practice of other established professions such as law and consider the differences and similarities in the way these two practice communities 'walk', 'talk' and 'think' their practices.

> Practices are bundles of sayings and doings that have existence beyond the particular individuals engaged in them. The ways practitioners speak about what they do and the actions in which they engage are not matters of individual choice but are an intrinsic feature of the practice itself. Practices connect material conditions with people and with work. They cannot be thought of separately from the conditions in which they exist – abstracting a practice from its context is to no longer have a practice. (Boud, 2016, p. 160)

Boud's view of practice draws our attention to the way practice (or work) is part of systems and cultures which in turn frame the expectations and practices (including language, choices, actions and conditions) of that work. This chapter provides an exploration of what practice is, what it is like and how it is influenced by the context(s) in which it operates.

PRACTICE: AN ENTRY POINT

Practice is essentially a socially constructed phenomenon, and a transactional social process that involves experience and action to inform human conduct. At its most fundamental level practice is what people do (DOING). But this is too simple, for what we do is influenced by who we are, our identity and personal frame of reference and where we are located (our context): we can call this our BEING, with the emphasis on being in the world or being engaged in the world. Heidegger (1926/1990) uses the term *Dasein* to refer to being-in-the-world in comparison to *Sein* (simply meaning being).

A critical aspect of being is that it is neither static, nor context free, thus we need to acknowledge the importance of the person's BECOMING, not only reflecting the way they change in response to, or in advance of, contextual influences but also the changes in the way they practise reflecting on how they are evolving. Doing as a deliberate act of practice is not possible without knowledge of self, of what we are doing and why. Our tacit and embodied doings are framed by our personal and occupational frames of reference even if we are not conscious of these at the time.

KNOWING is part of practice and encompasses multiple ways of knowing, including knowing arising through and from reflective experience, theorising, conceptualising and embodiment. These ways of knowing produce multiple forms of knowledge including propositional (research and theory-based) knowledge and non-propositional knowledge (knowledge derived from personal and professional experiences) (Higgs & Titchen, 1995).

Thus, practice can be interpreted as a combination of *doing, knowing, being and becoming* to pursue purposeful activities occurring within the social relationships of the practice context, the discourse of the practice and practice-system, and the settings that comprise the practice world (Higgs, 1999).

The value of conceptualising professional practice, particularly as comprising knowing, doing, being and becoming, is profound. These seemingly simple words take us to the depth of practice and practice ownership which deals with being, embodying and becoming a professional practitioner and owning, valuing and morally enacting the type of practice that professionals espouse and wish to provide to their clients. It also enables us to recognise and understand practising as a journey of growth in knowledge and capability, including the ability to work in unknown and unpredictable circumstances, and critical evolution of self-determined practice and career management capabilities, rather than simply doing a job in an externally controlled, legislated and legitimated environment.

In being and becoming the practitioner, the person and the professional merge with the goal of incorporating the following key dimensions of advanced professional practice (Higgs, 2016, 2019; Higgs, Richardson, & Abrandt Dahlgren, 2004) into their practice model and approach:

– *Professionalism*: Embodying a commitment to adopting a moral and ethical approach to practice, being critical (self-appraising), demonstrating duty of care.

- *Practice epistemology*: Understanding the way knowledge is determined and created within that practice world; taking an epistemological stance in one's practice.
- *Practice ontology*: Adopting a worldview of practice; having an understanding of what really exists in practice.
- *Epistemic fluency*: Exhibiting flexibility and adeptness in relation to different ways of knowing for and about the world.
- *Authenticity*: Embodying the capacity to be true to self, capable in practice, and genuine professionalism.
- *Practice wisdom*: Realising an embodied state of being that imbues and guides insightful and quality practice, and comprises self-knowledge, action capacity, deep understanding of practice and an appreciation of others.
- *Professional judgement:* Demonstrating the capacity to make highly skilled judgements and decisions that are optimal for the given circumstances of the client and the context. Such judgements are grounded in the unique knowledge base, frame of reference, self-knowledge, metacognitive ability, wisdom and reasoning capacity of the practitioner in the task of solving complex practice problems, such as demanding, moral and ethical issues; questions of value, belief and assumptions; and the intricacies of personal issues as they impact on people's lives. According to Nerland (2016), the legitimacy and trust of professions "rest on the capacity of practitioners to accomplish tasks and exercise professional judgment in ways that are informed, guided by, and validated against shared knowledge and conventions for practice" (p. 127).
- *Professional practice capabilities:* Having and using capabilities to practise ethically, soundly (credibly, justifiably, capably) and reflexively across each of the practice dimensions represented by doing, knowing, being and becoming in known/unknown, familiar/unfamiliar and current/potential practice spaces.
- *Professional language and interpersonal capacities and dispositions:* Being adept in the use of the range of communication and interaction abilities needed for practice including technical and professional discourse, talking and interacting with clients with respect, empathy, cultural fluency and humility (not just cultural competence), professionalism and acknowledgement of their own expertise and self-knowledge as part of their participation in the professional engagement.
- *Employability*: Beyond entry into and having the ability to work in the professional role, employability involves inhabiting and realising the role in ways that pursue quality and fulfilment plus ongoing practice advancement.
- *Practice collaboration*: Being effective, engaged, respectful, professional and collaborative in working within and across communities of practice.

Practice and Context

> [Practice is] doing, but not just doing in and of itself. It is doing in historical and social context that gives structure and meaning to what people do. In this sense, practice is always social practice. (Wenger, 1998, p. 47)

To implement practice, we need to understand it in context. Both individuals and social groups, in developing their own understandings of their worlds and shaping their (practice) actions, are influenced by their backgrounds: culture, language, history, gender, education and learning. From these backgrounds we develop beliefs, attitudes, dispositions, perspectives and standpoints. Gadamer (1997) used the term *horizon(s)* to refer to standpoints and perspectives. This term is useful as it enables us to conceptualise having a standpoint as reflecting what can be seen from a particular vantage point and recognising that this view is influenced by that position of viewing. This also opens the possibility, and practice, of inquiring beyond our immediate surroundings and understandings. Such inquiry requires an active fusion of horizons with other perspectives. Using these ideas, it is valuable to acknowledge that we all exist in situations where horizons are dynamic, open and shared or overlapping as opposed to fixed, absolute or unique; this is an important part of pursuing the social construction of knowledge and practice. This is not to say that we don't need a level of certainty in practice, and indeed things like codes of ethical conduct, standards and patient charters serve this purpose. However, even these change as practice evolves. In many instances practice is a collaboration or an interpretation or a set of options; it is important to recognise these changes and different perspectives in the pursuit and enhancement of quality practice and service to clients.

It follows that we need ways of understanding and paying attention to other people's motivations and what drives our practice communities Habermas (1968/1972) contended that ideas shape our interests (motivations) and actions; he divided interests into three categories: technical, practical and emancipatory. He argued that technical interest has a scientific bias and aims for technical success, and practical interest has a pragmatic bias and aims for consensual understanding, whereas emancipatory interest is directed towards critique and emancipation, and aims for critical understanding. In this way our interests influence our practice (the way we go about things) and our practices (the things we do) and our practice paradigms (philosophical and practice frameworks) and the practice communities we work within and, in turn, influence.

This influence from our education and practice communities shapes who we are as practitioners and how we practice. According to Bourdieu (1972/1977), our habitus (our system of embodied dispositions or tendencies created by an interplay between social structures and freewill) organises the way we perceive and react to our social world. Habitus can be viewed as the physical embodiment of cultural capital, including our deeply ingrained dispositions, habits and abilities that we have gained from our life experiences.

Using Bourdieu's notion, Wacquant (2005) describes habitus as "the way society becomes deposited in persons in the form of lasting dispositions, or trained capacities and structured propensities to think, feel and act in determinant ways, which then guide them" (p. 316).

Summary

This discussion so far indicates that we are influenced in our practices by:

- our background
- our evolution as interpretive beings against the backdrop of our situatedness
- our interests – both personal and community-based
- our perspectives and our capacity to engage in a fusion of horizons to expand our own horizons and understanding
- our embodied dispositions (habitus) in interaction with our social world.

PRACTICE THEORY

The scholarship around practice provides deep insights into ways we can understand its complexities. The significance of this work is highlighted by Schatzki (2001) who speaks of the "practice turn". This turn has prompted researchers and scholars to examine all kinds of activity through the lens of practice. Practice theory places the contextual analysis of practices as its main concern, rather than attributes or qualities of people (Boud, 2016).

Practice theory has been strongly influenced by the work of Heidegger and Wittgenstein. It is derived from various fields of study, particularly anthropology, sociology and history, that provide interpretations of how people with their diversity of intentions, motives and actions, pursue dynamic relationships and engagements with structures of society (often referred to as "the system") in order to shape the world they live in.

In the practice theory world, a key definition of practice is provided by Schatzki (2001); practice refers to "embodied, materially mediated arrays of human activity centrally organized around shared practical understanding" (p. 2). Common perspectives presented in practice theory (Schatzki, 2012) are that:

- practice is a social phenomenon embracing multiple people, comprising an organised constellation of diverse people's activities

- human life needs to be understood as forms of human practices that are the organised activities of multiple people

- human activity relies on something that cannot be put into words (building on Wittgenstein's *knowing how to go on*, Ryle's *know-how*, Bourdieu's *habitus*, Giddens' *practical consciousness* and Merleau-Ponty's *habits/schemas*).

According to Nicolini (2012) the value of practice theory lies in taking a practice-based view of social and human phenomena which allows us to take "radical departure from the traditional ways of understanding social and organizational matters" (p. 6). A distinctive practice-based approach, he contends:

- "emphasizes that behind all the apparently durable features of our world — from queues to formal organizations — there is some type of productive and reproductive work. In so doing it transforms the way in which we conceive of social order and conceptualize the apparent stability of the social world (the nature of social structures, in sociological jargon, as a socio-material accomplishment).
- forces us to rethink the role of agents and individuals; such as managers, the managed.
- foregrounds the importance of the body and objects in social affairs.
- sheds new light on the nature of knowledge and discourse.
- reaffirms the centrality of interests and power in everything we do". (p. 6)

PROFESSIONAL PRACTICE

Across this book we focus much of our discussion on occupations (jobs or work roles for which people are paid), particularly the sub-category of professional occupations. Professions are self-regulated occupational groups that have a body of knowledge and recognised role in serving society. They are accountable, under continual scrutiny and development, tertiary educated and are guided by codes of ethical conduct that are the foundation for practice decisions and actions by members of the profession (Higgs, 2018). Professions can be viewed as knowledge-based occupational groups that society entrusts to provide it vital services (Abbott, 1988; Freidson, 2001).

The term *professional* practice refers broadly to "the enactment of the role of a profession or occupational group in serving or contributing to society" (Higgs, McAllister, & Whiteford, 2009, p. 108). The practice of established occupations share a number of characteristics as indicated in the following quote.

> [Occupational and Professional] Practice ... encompasses the various practices that comprise occupations, be they professions, disciplines, vocations or occupations. For doctors, engineers, historians, priests, physicists, musicians, carpenters and many other occupational groups, practice refers to the activities, models, norms, language, discourse, ways of knowing and thinking, technical capacities, knowledge, identities, philosophies and other sociocultural practices that collectively comprise their particular occupation. (Higgs, 2012, p. 3)

Members of a profession are expected to complete an appropriate (commonly degree-based) intensive educational program. Such education encompasses the attainment of profession-specific knowledge, abilities and dispositions as well as graduate and generic abilities relevant to the 21st century and the expected attributes of professionalism through professional education and socialisation.

Professional socialisation refers to the acculturation process (through entry education, reflection, professional development and engagement in professional work interactions) by which individuals develop both the expected capabilities of the profession and a sense of professional identity and responsibility. (Higgs et al., 2009, p. 108)

According to Weidman et al. (2001) there are four stages of the process of engaging in socialisation. The stages reflect varied levels of understanding of and commitment to the professional roles of the future graduate. These stages are:

- *Anticipatory*. In this stage the newcomer is recruited into the learning program and becomes aware of the expectations held of a role incumbent. They hold preconceptions and stereotypical views of the role.
- *Formal*. In this stage the novice still holds idealised role expectations. They begin to receive formal instruction and observe experienced students and role incumbents. They transition to holding normative role expectations and to practise rather than observe the role. And they become veteran newcomers.
- *Informal*. In this stage the novice learns about the informal role expectations through immersion in their new culture. Much of the learning in this stage occurs within and through student cohorts and support groups. They begin to leave aside their student identity and develop a more professional identity.
- *Personal*. In this stage the role is internalised as individual and social personalities, roles and social structures become fused. The students now assess what success in the professional world means, what their career marketability is and what their plans for their professional directions and development might be.

Facing future practice requires a blend of professional education with education for future practice. Barnett (2004), for example, asks: "What is it to learn for an unknown future? It might be said that the future has always been unknown but our opening question surely takes on a new pedagogical challenge if not urgency in the contemporary age" (p. 247), the age of supercomplexity.

Evolution of the Professions

Kanes (2010) examined the challenges to professions and professionalism in the current context of major economic and social change characterised by complex market influences on professional services, insistent demands for accountability, increasingly rigid and limiting external regulation, and escalating ethical expectations. He identified a decrease in the significance of the traditional notions of autonomy, self-regulation and mastery with an increase in new trends such as multidisciplinary practice by boundary-crossing experts. His vision of a way forward draws on a UK medical report (Royal College of Physicians, 2005) that advocated for "a double transcendence: A transcendence of public over private interests, and the transcendence of inter-personal values of care over knowledge interests owned by practitioners" (cited in Kanes, 2010, p. 184).

Leicht (2015) also examined the future of the professions in the face of current challenges, particularly "three broad-sweeping social and cultural forces—market fundamentalism, the faith-like belief that unfettered free markets increase social and individual well-being and cultural fragmentation and post-modern skepticism that questions the commitments and values of the professions to universal conceptions of social progress, objectivity, and truth" (p. 11). He concluded that the continuation of the professions (in a modified form) will occur and will demand their capacity to adapt to cultural and market changes.

One such emerging adaptation in the professions is the introduction of blended professionals. They "not only cross internal and external institutional boundaries, but also contribute to the development of new forms of third space between professional and academic domains" (Whitchurch, 2009, p. 407). The theme of the future of the professions is explored further in Chapter 6 of this book.

Professionalism

> Professional behaviour (or professionalism) comprises those actions, standards and considerations of ethical and humanistic conduct expected by society and by professional associations and members of professions. (Higgs et al., 2009, p. 108)

Some occupations are recognised professions, some are emerging (becoming) professions (e.g. policing – see Green, 2015), some adopt the label of professional (e.g. athlete, builder) without recognition by society as a profession, some are neither seeking nor categorised as professions. The two key elements central to the claim for professionalism are enhancing the quality of service (Hoyle, 2001) and duty of care to the client. For Evetts (2014) professionalism is seen "as a special means of organizing work and controlling workers and in contrast to the hierarchical, bureaucratic and managerial controls of industrial and commercial organizations" (p. 778).

Evans (2008) speaks of new and modified professionalisms, using this term to reflect different ways of interpreting the quality of practice. She argues that power is a major feature in changing professionalism or de-professionalism, with a shift from autonomy to accountability (Hoyle & Wallace, 2005) while Ozga (1995) emphasises the importance of context, particularly policy context, in interpreting professionalism.

Both of these issues indicate a trend of external demands for and measures of quality replacing self-regulation. Perhaps this change reflects a failure of professions and professionals to deliver consistent quality to the public or perhaps it is part of the global trend to impose demands for accountability through external regulation.

Practice Development

This term refers to the development of a field of practice, such as occurs during the professionalisation of an occupation and the ongoing pursuit of improved quality, knowledge and advancement of the practices of the practice community. Education plays a critical role in professionalism and practice (Bradbury et al., 2015). Any field or occupation is evolving.

Many forces (society expectation, governmental policies, business market competitiveness, professional obligations, personal creativity and pursuit of learning) promote a striving for improvement, quality assurance and evolution of the practice of occupations. Part of professional expectations, expressed in practice codes of conduct, is the responsibility to contribute to the advancement of practice in and of the profession. It is this which we call practice development.

Nerland (2016) presents a powerful case for the diligent continuing development of the professions. She argues that professions can be seen as distinct knowledge cultures that are constituted by a set of knowledge practices that collectively define expertise in that profession and that distinguish professional practitioners from actors outside the profession (Nerland & Jensen, 2014). Through these knowledge practices: knowledge and practice are stabilised and developed, newcomers are welcomed into the profession, the profession maintains its legitimacy and sustainability.

A key challenge for the professional, argues Nerland (2016) "is that both knowledge and established conventions are generally questioned. Thus, the sustainability of professional practices requires continuous development and re-interpretations from practitioners" (p. 127).

Adopting a range of strategies for practice development is valuable; these can include (Higgs & Titchen, 2001b, pp. 530-531):

- Developing a greater understanding of professional knowledge which involves taking a critical look at what we call knowledge, what knowledge is accepted (challenged and unchallenged) in our professions and how this knowledge relates to and arises from our practice (see Eraut 1985, 1994; Higgs & Titchen, 2001a).

- Making knowledge derived from practice-based experience (professional craft knowledge) more explicit. Professionals can seek to become more attentive to and aware of how they practise and then test this practice knowledge and if found to be credible they can integrate it with existing knowledge.

- Exposing practice knowledge claims to public scrutiny. Consensual validation of this new knowledge results in the practice knowledge of the individual evolving to become the practice knowledge of the group and/or profession.

- Generating practice-based theory both as input to and a result of research. These processes and theory development promote critique and debate of both professional and practice-based knowledge.

- Re-mapping educational curricula to include new knowledge and practices. Barriers to overcome in this re-mapping include: (a) Lack of recognition of the importance of personal development as part of professional education curricula, (b) Over-emphasis on propositional knowledge in teaching and assessment, making professional craft knowledge and personal knowledge seem less relevant and credible, and (c) Modelling of professional behaviours which are no longer considered appropriate such as professional depersonalisation (Conroy, 2001).
- Review of the nature of professional practice.
- Wider structural and cultural changes and reconfiguration of professionalisation, professional identity and socialisation to promote practice transformation using critical social science frameworks.

PRACTICE AND WORK IN CONTEXT

In addition to the expanse of literature and theory concerning practice introduced above, we can also conceive of practice as work. Here we are referring to work as paid occupation. Both work and occupational/professional practice, as discussed above, are social constructs and socio-historically and culturally shaped sets of activities and roles in society. With these ideas, practices and social frames of reference as a backdrop, employability in action cannot be regarded as a fixed idea or set of requirements/dimensions. Consider, for instance, the following aspects of work: career instability or flexibility and work security.

Progressively within the past 30 years, for instance, and exponentially in the past 10 years, work has become increasingly less stable and more vulnerable. A significant part of this is due to massive contextual changes such as globalisation and technological changes – in particular the rise of automation and artificial intelligence, information and communications technological changes and pervasiveness, environmental changes and major population movements. Each of these changes have had progressive and significant impacts on work – its availability, accessibility, flexibility and stability.

People who are at the influence/management/political end of the labour market as well as people immersed and perhaps drowning in the volatility of today's labour market, are both concerned about the growth of vulnerable work (International Labour Organization, 2017). Workers who are particularly vulnerable (young and older people, those without required job experience, less well-educated people, people with high levels of family responsibilities that restrict their work choices, those with disabilities, those facing socioeconomic poverty) suffer most. People in popular graduate occupations also face vulnerable work conditions due to fierce competition for work and networked forms of employment, in the face of rising numbers of graduates and declining numbers of traditional full-time positions. Vulnerable workers are more likely to experience poor working conditions, limited access to social protections, volatile income, little career development and greater exposure to unethical behaviours including bullying and harassment (see Hajkowicz, Cook, & Littleboy, 2012; Hooley, Sultana, & Thomsen, 2017).

Work and Self

What is it that individuals can do to optimise their future work and labour market options? Many answers can be presented for this question and are discussed throughout this book. A few insights are introduced here. The idea and strategy of deep work is proposed by Newport (2016). In a world of distraction, deep work refers to "professional activities performed in a state of distraction-free concentration that push your cognitive capabilities to their limit. These efforts create new value, improve your skill, and are hard to replicate" (p. 3). Deep work is critical in wringing every "last drop" of value from the individual's current intellectual capacity. Newport compares deep work, with sharp contrast, to the typical behaviour of most modern knowledge workers who spend much time in shallow, busy work. A key reason for this is the focus on network tools (e.g. social media) and the way they dominate and fragment knowledge workers' attention, taking time away from deep work. Catch cries such as "the tyranny of email" typify this dilemma.

Gardner (2008) proposes that there are 5 minds that are "particularly at a premium in the world of today and will be even more so in the future" (p. 4). They span both cognitive and human spectrums and are broad uses of the mind that use but are distinct from the human intelligences. The 5 minds are:

- *The disciplined mind.* This thinking refers not to subject matter (to be learned) but rather a distinctive disciplinary way of thinking about the world. For instance, the discipline of chemistry thinks and uses knowledge in a different way to the field of medicine or architecture.
- *The synthesizing mind.* This thinking draws together discrete or disparate elements in more manageable syntheses such as narratives, taxonomies, complex concepts, rules/aphorisms, metaphors/images/themes, embodiments without words, theories and metatheories.
- *The creating mind.* This creativity arises from interaction between three autonomous elements: the individual, the cultural domain and the social field.
- *The respectful mind.* Fulfilling our roles as citizens in society, including at work, requires respect – for others, for differences, etc.
- *The ethical mind.* Ethics deals with rights, obligations and responsibilities and doing good, in work and social life.

These and other examples provide a means for developing self to meet the supercomplexity demands of future practice. Other strategies to address such demands include framing, agency, reflexivity, deliberateness, re-claiming social purpose, resilience, re-imagining, risk taking and educational innovations. (See also Chapters 2, 6, 13, 14, 18.)

CONCLUSION

To pursue practice futures requires us to first interpret practice; in this chapter I interpret practice as comprising doing, knowing, being and becoming. Practice exists always in context, including the personal, social, occupation, local, organisational, historical, geographical, national and political contexts. Professional practice likewise operates within a multidimensional, dynamic context that encompasses recognising the nature and challenges of professions and their future possibilities, the critical demands of professionalism, the embodiment and challenges of professional work, and the external political and accountability contexts of the professions and their evolution. Education and professional development are critical factors in exploring and enacting practice relevant to context.

REFERENCES

Abbott, A. (1988). *The system of professions*. Chicago, IL: University of Chicago Press.

Barnett, R. (2004). Learning for an unknown future. *Higher Education Research & Development, 23*(3), 247-260.

Boud, D. (2016). Taking professional practice seriously: Implications for deliberate course design. In F. Trede & C. McEwen (Eds.), *Educating the deliberate professional: Preparing for future practices* (pp. 157-173). Switzerland: Springer.

Bourdieu, P. (1972/1977). *Outline of a theory of practice* (R. Nice, Trans.). Cambridge, England: Cambridge University Press.

Bowden, J., & Marton, F. (1998). *The university of learning: Beyond quality and competence in higher education*. London, England: Kogan Page.

Bradbury, H., Kilminster, S., O'Rourke, R., & Zukas, M. (2015). Professionalism and practice: Critical understandings of professional learning and education. *Studies in Continuing Education, 37*(2), 125-130.

Conroy, S. (2001). Professional craft knowledge and curricula: What are we really teaching? In J. Higgs & A. Titchen (Eds.), *Practice knowledge and expertise in the health professions* (pp. 178-185). Oxford, England: Butterworth-Heinemann.

Eraut, M. (1985). Knowledge creation and knowledge use in professional contexts. *Studies in Higher Education, 10*(2), 117-133.

Eraut, M. (1994). *Developing professional knowledge and competence*. London, England: Falmer Press.

Evans, L. (2008). Professionalism, professionality and the development of education professionals. *British Journal of Educational Studies, 56*(1), 20-38.

Evetts, J. (2014). The concept of professionalism: Professional work, professional practice and learning. In S. Billett, C. Harteis, & H. Gruber (Eds.), *International handbook of research in professional and practice-based learning* (Vol. 1, pp. 29-56). Dordecht, The Netherlands: Springer.

Fenwick, T., Nerland, M., & Jensen, K. (2012). Sociomaterial approaches to conceptualising professional learning and practice. *Journal of Education and Work, 25*(1), 1-13.

Freidson, E. (2001). *Professionalism: The third logic*. London, England: Polity Press.

Gadamer, H.-G. (1997). *Truth and method*. New York, NY: Continuum.

Gardner, H. (2008). *5 minds for the future*. Boston, MA: Harvard Business Press.

Green, B. (2009). Introduction: Understanding and researching professional practice. In B. Green (Ed.), *Understanding and researching professional practice* (pp. 1-18). Rotterdam, The Netherlands: Sense.

Green, T. (2015). *Becoming a police academic: From practitioner to educator* (Unpublished doctoral dissertation). Charles Sturt University, Australia.

Habermas, J. (1968/1972). *Knowledge and human interest*. (J. J. Shapiro, Trans.). London, England: Heinemann.

Hajkowicz, S., Cook, H., & Littleboy, A. (2012). *Our future world: Global megatrends that will change the way we live*. Sydney, Australia: CSIRO.

Heidegger, M. (1926/1990). *Being and time* (J. Macquarrie & E. Robinson, Trans.). Oxford, England: Basil Blackwell.

Higgs, J. (1999, September). *Doing, knowing, being and becoming in professional practice*. Presented at the Master of Teaching Post Internship Conference, The University of Sydney, Australia.

Higgs, J. (2012). Practice-based education: The practice-education-context-quality nexus. In J. Higgs, R. Barnett, S. Billett, M. Hutchings, & F. Trede (Eds.), *Practice-based education: Perspectives and strategies* (pp. 3-12). Rotterdam, The Netherlands: Sense.

Higgs, J. (2016). Practice wisdom and wise practice: Dancing between the core and the margins of practice discourse and lived practice. In J. Higgs & F. Trede (Eds.), *Professional practice discourse marginalia* (pp. 65-72). Rotterdam, The Netherlands: Sense.

Higgs, J. (2018). Judgment and reasoning in professional contexts. In P. Lanzer (Ed.), *Catheter-based cardiovascular interventions: Knowledge-based approach* (2nd ed., pp. 15-25). Cham, Switzerland: Springer International.

Higgs, J. (2019). Re-interpreting clinical reasoning: A model of encultured decision making practice capabilities. In J. Higgs, G. Jensen, S. Loftus, & N. Christensen (Eds.), *Clinical reasoning in the health professions* (4th ed., pp. 13-31). Edinburgh, Scotland: Elsevier.

Higgs, J., McAllister, L., & Whiteford, G. (2009). The practice and praxis of professional decision making. In B. Green (Ed.), *Understanding and researching professional practice* (pp. 101-120). Rotterdam, The Netherlands: Sense.

Higgs, J., Richardson, B., & Abrandt Dahlgren, M. (Eds.). (2004). *Developing practice knowledge for health professionals*. Oxford, England: Butterworth-Heinemann.

Higgs, J., & Titchen, A. (1995). The nature, generation and verification of knowledge. *Physiotherapy, 81*, 521-530.

Higgs, J., & Titchen, A. (Eds.). (2001a). *Practice knowledge and expertise in the health professions*. Oxford, England: Butterworth-Heinemann.

Higgs, J., & Titchen, A. (2001b). Rethinking the practice-knowledge interface in an uncertain world. *British Journal of Occupational Therapy, 64*(11), 526-533.

Hooley, T., Sultana, R., & Thomsen, R. (Eds.). (2017). *Career guidance for social justice: Contesting neoliberalism* (Book 16). New York, NY: Routledge.

Hoyle, E. (2001). Teaching: Prestige, status and esteem. *Educational Management & Administration, 29*(2), 139-152.

Hoyle, E., & Wallace, M. (2005). *Educational leadership: Ambiguity, professionals and managerialism*. London, England: Sage.

International Labour Organization (ILO). (2017). *World employment and social outlook: Trends 2017*. Geneva, Switzerland: International Labour Office.

Kanes, C. (2010). Studies in the theory and practice of professionalism: Ways forward. In C. Kanes (Ed.), *Elaborating professionalism: Studies in practice and theory* (pp. 183-198). New York, NY: Springer.

Leicht, K. T. (2015). Market fundamentalism, cultural fragmentation, post-modern skepticism, and the future of professional work. *Journal of Professions and Organization, 0*, 1-15.

Nerland, M. (2016). Learning to master profession-specific knowledge practices: A prerequisite for the deliberate professional? In F. Trede & C. McEwen (Eds.), *Educating the deliberate professional: Preparing for future practices* (pp. 127-139). Switzerland: Springer.

Nerland, M., & Jensen, K. (2014). Changing cultures of knowledge and professional learning. In S. Billett, C. Harteis, & H. Gruber (Eds.), *International handbook of research in professional and practice-based learning* (pp. 611-640). Dordrecht, The Netherlands: Springer.

Newport, C. (2016). *Deep work: Rules for focused success in a distracted world*. London, England: Piatkus.

Nicolini, D. (2012). *Practice theory, work and organisation: An introduction*. Oxford, England: Oxford University Press.

Ozga, J. (1995). Deskilling a profession: Professionalism, deprofessionalisation and the new managerialism. In H. Busher & R. Saran (Eds.), *Managing teachers as professionals in schools* (pp. 21-37). London, England: Kogan Page.

Royal College of Physicians. (2005). *Doctors in society: Medical professionalism in a changing world* (Report of a Working Party of the Royal College of Physicians of London). Retrieved from https://shop.rcplondon.ac.uk/products/doctors-in-society-medical-professionalism-in-a-changing-world?variant=6337443013

Schatzki, T. R. (2001). Introduction: Practice theory. In T. R. Schatzki, C. Knorr, & E. von Savigny (Eds.), *The practice turn in contemporary theory* (pp. 1-14). London, England: Routledge.

Schatzki, T. R. (2012). A primer on practices: Theory and research. In J. Higgs, R. Barnett, S. Billett, M. Hutchings, & F. Trede (Eds.), *Practice-based education: Perspectives and strategies* (pp. 13-26). Rotterdam, The Netherlands: Sense.

Wacquant, L. (2005). Habitus. In J. Beckert & M. Zafirovski (Eds.), *International encyclopedia of economic sociology* (pp. 317-320). London, England: Routledge.

Weidman, J. C., Twale, D. J. & Stein, E. L. (2001). *Socialisation of graduate and professional students in higher education: A perilous passage?* (ASHE-ERIC Higher Education Report Vol. 28, No. 3). New York, NY: John Wiley and Sons, Inc.

Wenger, E. (1998). *Communities of practice: Learning, meaning and identity*. Cambridge, England: Cambridge University Press.

Whitchurch, C. (2009). The rise of the blended professional in higher education: A comparison between the United Kingdom, Australia and the United States. *Higher Education, 58*(3), 407-418.

Joy Higgs AM, PhD (ORCID: https://orcid.org/0000-0002-8545-1016)
Emeritus Professor, Charles Sturt University, Australia
Director, Education, Practice and Employability Network, Australia

STEVEN CORK AND DEBBIE HORSFALL

2. THINKING THE UNTHINKABLE

*Challenges of Imagining and Engaging with
Unimaginable Practice Futures*

THE CHALLENGES ARE SUBSTANTIAL

What might practice and work look like in a range of possible futures? This is the central question posed by this book. It seems a straightforward question that simply requires some anticipation of how the world might change and what roles practice might play in different future worlds. But asking this seemingly simple question causes us to confront some powerful limitations of human thinking processes. In some ways, it challenges us – both the authors of this book and its readers, as members of societies that are creating, and preparing for, possible futures – to think in ways that we normally find *unthinkable*. In this chapter, we will consider evidence suggesting that human brains have evolved powerful ways to avoid engaging with complex and uncertain questions like "What might the future be like?" We will consider many implications of this avoidance, including powerful ideas that go largely unchallenged and so leave humanity vulnerable to *inevitable surprises*, many of which will be undesirable and even potentially catastrophic (Schwartz, 2003). We will touch on some aspects of societal governance that inhibit thinking about unthinkable possibilities, but we will also give examples of processes by which groups of people can help, and are helping, one another to break free of thinking constraints to make us better prepared for what the future might hold. And, of course, we will discuss what all this might mean for our focal question about the possible futures of practice.

ASKING THE RIGHT QUESTIONS

The phrase "thinking the unthinkable" has been used by practitioners of *strategic foresight* (the discipline of thinking systematically about multiple plausible futures) to emphasise the reality that humans appear to have a strong need to believe that the future is predictable and so find it *unthinkable* to imagine futures in which things don't happen in ways that have been seen before. Nicholas Taleb illustrated this with his book *Black Swan* (Taleb, 2007). He reports that the first black swan taken to England from Australia was treated as a fake, because scientists in England had never seen a black swan before (European swans are white) and their default mode of thinking was that something never seen, or never predicted, before must be treated as highly improbable.

We will discuss human psychology and the bases for thinking constraints in more detail, below. In the meantime it is helpful to understand that thinking constraints most often lead us to focus on *thinkable* questions (i.e. ones we are used to asking). So, for most of us, when we see a question like "What might the future of practice be like?" we straight away expect a simple answer, based on current understanding of the word practice and the views of experts who, we readily believe, can tell us how things might change and what that change will mean for practice.

But, if we stop and think a bit more deeply, we soon realise that this seemingly simple question begs many others. Is practice understood in the same way across society? Might practice as currently conceived exist in all possible futures? What futures might society be interested in: possible, probable, preferable, official, feared…? Do all members of society have a clear idea of what sort of future they would prefer? Do all members of society prefer the same future? What about people who live in societies different from ours, do they have the same questions? Do the so-called powerless ask different questions to those with power and influence? If preferences differ, whose preference should prevail if we decide to shape the future? Who should be engaged in addressing these questions?

Should society seek immediate answers to these questions or is it more important to generate dialogue about possible answers, knowing that answers might only emerge after a long period of asking them? Is it the answer that society should be seeking or is it the conversations involved in exploring possible answers, which might help societies understand the different ways in which their members make sense of the world? And perhaps the two deepest questions are: firstly, why should we care about the possible futures of practice in the first place; and, secondly, who should be interested (e.g. is it those who educate and train practitioners, or practitioners and/or their professional bodies, or is it broader, perhaps even the whole of society, who should be interested)?

In the following sections we explore what we (the authors of this book) mean by practice and why we think it is important to ask about its possible futures. Then we review evidence about why it is hard for humans to think in unconstrained ways about this type of issue, before suggesting some ways in which human thinking limitations can be engaged with to enable at least some *thinking of the unthinkable*.

WHAT DO WE MEAN BY PRACTICE?

In Chapter 1 of this volume, Higgs discusses how we, the authors, interpret *practice* and why we choose to focus especially on *professional practice*. She explains that practice is about *doing, being, becoming* and *knowing*. It is partly what people in occupations do, but it also is influenced by who we are and the many contexts within which we exist, and it changes as people evolve and as people develop multiple knowledge sets that help them understand and interact with the world in many different ways.

It would be easy to interpret the question "What might the futures of practice be like?" as "How might different futures affect the ability of professions to keep practising as they currently do?" But this would be a narrow interpretation. Instead, we could consider professional practice to be more broadly about helping society negotiate complex information and concepts, for the common good. Within this interpretation we acknowledge that, professionals and societies evolve, neither is static. As such, people develop new expectations and norms and have different sets of knowledge and tools to collect and use that knowledge at their disposal. So, in the future, what we currently think of as professional practice might look and be very different in different possible futures.

As we discuss throughout the rest of this chapter, it is hard for humans in general (including us, the authors of this chapter) to think about the full range of possibilities for practice in the future because our brains resist – for understandable reasons – imagining situations that we have not seen, or heard about, before.

HUMAN PSYCHOLOGY AND THINKING LIMITATIONS

We cannot hope to review the immense literature on how humans think. Instead, we review selected sources that capture the key elements of that literature in relation to the points we are making in this chapter.

Many researchers have demonstrated the tendency of humans to filter information; to hear and selectively recall information that fits with the knowledge sets that they use to make sense of the world (Ariely, 2008; Czechmihaley, 2008; Roxburgh, 2003; Snowden, 2002). The metaphor of *mental models* has been evoked to explain this phenomenon (Jones et al., 2011). One idea that appears across much of the literature is that mental models are a mechanism that protects human brains from being overloaded by the vast amounts of information they potentially receive every moment of every day and allows humans to focus on the information they believe is most important at any point in time. For example, Daniel Gilbert suggests that during the time when human brains were evolving their cognitive abilities it was most important to focus on four types of threats: intentional, immoral, imminent and instantaneous (Gilbert, 2010). This, he argues, explains, for example, why humans react more strongly to terrorism (an intentional, immoral, imminent and instantaneous threat) than influenza, which kills many more people every year but which had none of the four characteristics listed above.

Psychologists and behavioural economists have demonstrated a wide range of decision-making practices that arise from humans taking shortcuts in their thinking, prioritising some values over others without sound evidence, and, essentially, acting on familiar patterns of events of information without thinking about whether those patterns are relevant to current situations (Ariely, 2008; Roxburgh, 2003; Snowden, 2002).

Those researching organisational learning and societal pathologies from a systems perspective argue that many failures of humans to anticipate or prepare for future challenges and opportunities result from a rush to simplify complexity by pattern matching, filtering information and other ways of avoiding engagement with that complexity (Senge, 2010; Snowden, 2002). As mentioned previously in this chapter, another pervasive symptom of the human need to simplify complexity without engaging with it, is the tendency of humans to imagine order and inevitability where it does not exist and to conclude that the future can be predicted from the past (Taleb, 2007).

THERE IS NO ALTERNATIVE (OR IS THERE?)

Constraints on human imagination have led to the normalisation of thinking across society about what is possible in the future and what it not. This normalisation is not always innocent, or benevolent. For example, various professions have exploited people's limited ability to weigh up evidence when making comparisons between alternatives. Much advertising is based on the inability of people to see illogical connections between idealised lifestyles and purchasing a promoted product. A key aspect of the marketing of items in shops is their placement relative to one another to create the impression of value for money. The selling of products as diverse as residential real-estate and magazine subscriptions is enhanced by the comparisons between offering or "deals", and a common strategy of groups wishing to influence public opinion is the create false choices that rely on people not considering the choice (Edis, 2015).

The acronym TINA (There Is No Alternative) has been used by futures thinkers to challenge those who think certain aspects of the future are set in stone. For example, in the late 1990s many business leaders thought that globalisation, liberalisation and technology would inevitably grow and shape the future. These were forces that "even governments cannot resist" (Royal Dutch Shell, 1995, p. 2). Since then, various futures-thinkers have questioned all of these TINAs, producing scenarios for plausible futures in which globalisation falters or even reverses, liberalisation becomes much less popular, and technology fails to deliver on expectations. Some of these sorts of scenarios are considered in Chapter 3 of this volume in relation to the future of work and practice. At the extreme, for example, the Global Scenarios Group[1] envisaged a divided "Fortress World" future on the one hand and on the other a "Great Transition" future in which the world rejects many of the values of market-driven societies to return to more sustainable ways of living (Raskin et al., 2002). These scenarios have been adapted in many subsequent futures-thinking exercises (Hunt et al., 2012). Similarly, Joanne Macy has popularised the great transitions style of thinking in her movie *The Great Turning* (Macy, 2018) and numerous books such as *Active Hope: How to Face the Mess We Are in Without Going Crazy* (Macy & Johnstone, 2012).

The concept of TINA has also been used by others to draw attention to false, or at least unchallenged, beliefs about unchangeable, often undesirable, aspects of society (Mies & Shiva, 1993; Shiva, 1993). For example, the feminist movement challenges the belief that Western society will always be patriarchal or that women can only play certain roles. The fact that the latter belief has been largely discounted shows how, with time, society's views about plausible futures can change and what is viewed as highly improbable at one point in time can become a realised future at a later time.

HOW SOCIETY MIGHT THINK THE UNTHINKABLE

The concept of TINAs is one mechanism that has been used to encourage thinking about the unthinkable (i.e. that which is unimaginable unless people are pushed to break free of their safe thinking space and thinking constraints). It is no surprise that TINAs arose from the discipline of strategic foresight, as this discipline is founded on the observations that preparing for future challenges and opportunities requires envisioning as broad a range of plausible (not necessarily likely) futures as possible, and that this is only possible if those doing the thinking are given support to think the unthinkable. Typically, a process to provide this support involves the creation of trust, reaching agreement to "suspend disbelief", acknowledgement and questioning of assumptions about the past, present and future, and encouragement of true dialogue that involves as much listening and understanding as it does talking and convincing (Scearce, Fulton, & Global Business Network Community, 2004).

One recent example of a process encouraging true dialogue about multiple futures was the *Australia 2050* project supported by the Australian Academy of Science. This project brought together around 50 leading thinkers from across Australian society, with a key selection criterion that invited participants had demonstrated a willingness to consider the views of others (Cork et al., 2015). The participants were asked to listen to one another, and understand one another's world views, before forming their ideas about a range of plausible futures. This workshop was intended to model a process that could be used by groups of people generally to engage in structured conversations about the future.

At a less focused level, science-fiction writing and other genres of public communication via the arts and humanities, that acknowledge and challenge assumptions while envisaging non-typical futures, play a role in helping society think the unthinkable. Much of Ursula Le Guinn's work falls into this category (e.g. Le Guinn, 2001). A feminist science fiction writer she often depicts future societies which eschew the typical dystopian fantasies of much science fiction where the baser values of society have often gained ascendance.

The *Earthsea* series[2] is such an example. *Star Trek* is another example, where the future is depicted as peaceful and harmonious and the purpose of being human is to create art, pursue mysteries, explore and develop knowledge. For many people, envisaging a society free from war and conflict, or fixed gender states, is currently unthinkable, with trite arguments of "humans are violent by nature" being trotted out. What Le Guinn and the writers of *Star Trek* do, via popular culture, is offer alternative futures for us to think about. Importantly they provide detailed imaginary visions of what this might look like, and the pictures are populated with people who look and sound like us in many ways, thus enabling these imagined alternative futures to resonate and even be embraced as possible, even desired, alternatives. In this way we can begin to think other-*wise*.

When we start thinking the unthinkable, our approach to considering the futures of practice is broadened to the limits of current human contemplation about the future of our species. As one example, we draw on a milestone publication in the development of feminist post-humanist theory, *A Cyborg Manifesto* (Haraway,

1991). Haraway observed that three boundaries have been blurred during the 20th Century: those between human and animal, animal-human and machine, and physical and non-physical. In an early challenge to dichotomous thinking, she challenges many of the labels and taxonomies that we apply to people, including self versus other, culture versus nature, male versus female, civilised versus primitive, right versus wrong, truth versus illusion, and God versus man, which, she suggests are artificial dualities that have been used to justify domination of some people by others. She suggests that one way to remove our biases is to see the world objectively, as cyborgs might. Thus, we would not place any special importance on the Garden of Eden as we would see it as dirt and vegetation, we would not see all women as the same and all men as the same – but different from women, and we would consider interrelationships between people in terms of real affinities, which would cut across currently perceived boundaries and categories of gender, race, politics, culture etc.

There has been a huge amount of debate about these ideas, but our reason for raising them is to illustrate an extreme possibility for the focus of this book: that we might in the future consider the concepts of practice and professions as arbitrary and artificial constructs of the 19th and 20th centuries and we might see no clear boundaries between people who can handle complex information and people who cannot, and we might not see a role for a particular group of people who society relies on to help it make sense of complex information. Instead, thinking, knowledge and expertise would be shared widely and AI would complement whatever attributes an individual brought to a social role, to create a melded human-machine unit capable of interpreting and applying all relevant knowledge to a particular role.

Attempting to inhabit the Cyborg's view can free our minds to think the unthinkable which is no easy task. For example, our thinking is bound by ideologies, belief systems and available language and discourse as eloquently illustrated by Margret Wertheim (1997) in the book *Pythagoras's Trousers*. Ancient mathematicians and astrologers faced a dilemma: the universe was created by God and God created perfection; the circle was seen as the perfect shape so all planet orbits had to therefore be circles. However, empirical evidence showed that planets' orbits were not circular. Due to the overarching belief systems of the times, and the severe punishments meted out for thinking outside of these systems, usually death – the problem was not solved for hundreds of years.

Embracing, co-opting or creating new ways of talking about things is another strategy for thinking differently. Here our minds turn to language and concepts offered by complexity science and critical pedagogies. Using language and concepts such as: acting other-wise or outside an economy of the same; the epistemology of emergence; and, creating the space of the not yet possible provides conversational spaces for us to imagine, create, innovate and think differently, beyond ideology perhaps? Further, ideas such as "the space of the not yet possible" suggests that there is something beyond the limits of our current thinking, imaginings and understandings.

Osberg (2009) argues that if we can already imagine something – future practice for example – then it already exists, if only in our minds. The challenge is to move further than the current limits of our imaginations by exploring the space of the impossible, or that which cannot yet be imagined as possible. Osberg (2009) and Biesta, Osberg, and Cilliers (2008) argue that this is only possible if we think in a non-linear way (i.e. not cause and effect) using emergentist logic which occurs in relationship with the world. This language offers us possibilities to think differently, to move towards thinking the unthinkable, or the not yet possible. And how are we to do this? Perhaps the people writing/thinking as the "disruptors" offer us a clue. Pascale, Sternin, and Sternin (2010) say we are more likely to act our way into new ways of thinking than think our way into new ways of acting.

THE POTENTIAL OF ARTIFICAL INTELLIGENCE

Here we find ourselves sliding into one of the futures considered in Chapter 3 of this volume (Cork & Alford). Susskind and Susskind (2015) suggested that the ultimate future for societies will be one in which information is freely available to all people, through the mediation of artificial intelligence (AI). They suggest that AI will make sense of information in ways that all members of society can understand and act upon. Furthermore, they suggest that AI might be in a far better position to apply, impartially, rules based on ethical and moral principles than human decision makers have or could.

In Chapter 3, Cork and Alford consider the possible roles of the human successors of today's professionals, but they also consider futures in which AI fails to deliver on expectations. Our point here is that there is an amazing diversity of possible ways in which society might organise itself and its knowledge in the future, with unclear boundaries between humans and technologies, ranging from no change from today to change that is *unthinkable* today.

DANGEROUS KNOWLEDGE

The sort of unthinkable thinking explored above, which we have barely scratched the surface of, is disruptive in many ways. According to many psychologists, ideas are unthinkable for good reasons. People feel very uncomfortable considering futures they cannot control or comprehend. We feel helpless, confused, even threatened.

Some of society's leaders and influencers fear the effects of such ideas becoming widespread. Some are motivated by concerns for the mental wellbeing of their people. After all, as Richard Eckersley observed in his research on young people's "fear of the Apocalypse" (Eckersley, 2008), when people feel concerned and uncertain about the future they might engage positively in activities to seek desirable futures, or they might disengage by seeking fundamentalist beliefs that offer simplified and reassuring views about the future and a person's role in it, or they might simply retreat to nihilism and live for the present with no regard for the future.

Other influencers see danger in ideas arising from thinking the unthinkable. These ideas could undermine established or hoped-for power and influence. The current cannibalisation (Baird, 2018) of the #metoo movement with three recent examples in Australia of women victims of sexual assault by powerful men, being outed by other powerful men, serve as potent examples of how imagining alternative futures can elicit a backlash from people who wish to maintain the patriarchal status quo.

Here the unthinkable world is one where violence is not sexualised, or used to keep women (and some men) in their place. The "danger" though is that patterns of behaviour, systems of domination and relations of power would have to change: the men (and some women) who are the beneficiaries of current systems and relations stand to lose, or be undermined. Naturally, such leaders and influencers feel threated. In order to minimise the threat, the apparatus of the state, (for example, schools, churches, family, media, trade unions and law, while formally outside state control serve to transmit the values of the state [see Althusser, 1971]) gets to work to keep our thinking in line with current dominant values, or the values which serve the interests of the ruling elites and influencers – in this example the values of capitalist patriarcy. These practices of hegemony mean that we, as a society, believe there are no alternatives to the current order, or that the alternatives are too difficult, or too implausible to realise.

FUTURE; WHICH FUTURE?

When most people talk about the future they emphasise *the* before the word *future*, indicating that they are thinking of one future. But futures thinkers have identified a range of futures that people think about. At any time, or in any particular situation, a person might have any one of these types of future in mind (probably without realising it).

Often, the future referred to in public discourse and the literature is the future generally considered as the most *probable.* Probability here is not established through any scientific process, because the future is inherently unpredictable in the long term and facts from the present and past are only helpful over short time frames in which the complexity of, and the uncertainty about, possible futures is within limits and/or ability to control the future is high (Peterson, Cumming, & Carpenter, 2003; Zurek & Henrichs, 2007). Research on how people in general determine the truth or likelihood of what they are told indicates that repetition is one of the strongest determinants (Weaver et al., 2007). Therefore, the *probable* future is the one that is most often repeated across society. It might or might not reflect a likely short-term future, but it almost certainly does not reflect the longer-term future, for all the reasons we have discussed in this chapter.

The *official* future is the one that governments and other authoritative institutions communicate to society. The *official* future is strongly influenced by political and other ideologies (especially economic ideologies in today's world) and might or might not resemble the *probable* future. For example, supporters of a ruling political party might see the *official* future as the *probable* future, whereas supporters of opposition parties, or those who are disenchanted with politics in general, are likely to see the *probable* future as quite different from the *official* future.

The *preferred* future is the one we desire. In our Western capitalist worlds, most people have only a vague idea of what their preferred future might be. They might be clear about some aspects (e.g. they would like to have a job and a good quality of life) but they have not though in detail about what a good life might be like or what sort of society and broader world might provide that life. Furthermore, they either believe that others want the same future or haven't thought about whether others want the same future. A recent research project developed a set of scenarios for the future of Australia (Costanza et al., 2015). These scenarios were then sent, with a survey, to around 300 Australians, who were asked a series of questions about their reactions to the scenarios, including which scenario they preferred and which they thought most likely to eventuate (Chambers et al., 2018, in press). There was a clear preference for one of these scenarios, but there were substantial numbers of people who voted for other scenarios, indicating that preferences varied and that it was not possible to talk about one future *preferred* by all Australians. The scenario that was preferred by most people was different from the one most thought to be likely, suggesting that many people think they have no influence over how the future unfolds or think that their preferences do not reflect those of the broader population that will determine how the future unfolds.

The *feared* future might be similar to the *probable* or the *official* future if one is pessimistic or it might be quite different if one is optimistic. As with the *preferred* future, the *feared* future might differ considerably between people. All of these types of future are present and are discussed across society, to varying degrees. The types of futures that professional futures thinkers focus on are the set of *plausible* futures, which includes all of the types of futures listed above. These are futures that explore the full range of what we currently believe to be within the bounds of possibility, bounded only by immutable laws of physics. Such futures are all highly improbable, because each is a subset of an infinite number of possibilities.

Insights into *plausible* futures can only emerge after systematic consideration of what factors drive change and how these might, *plausibly* if not *probably*, maintain or change their trajectories of change and/or interact with one another in the future. Obviously, given our previous discussion, achieving useful insights into any types of futures requires action to alleviate human thinking constraints. As discussed above, the processes involved in thinking about potential futures also serve to help people listen to, and understand, one another's assumptions, hopes and fears about the future. The greatest value in these processes is being involved. Unfortunately, people who have not been involved receive the ideas and conclusions without experiencing the challenging of their mental models and information filters that the interactive process provides.

CONCLUSION

Asking the question "What might the futures of practice be like?" is far from a straightforward question. In fact, it generates a host of other questions, some of which we have raised in this chapter. Answering these questions requires thinking the unthinkable – i.e. thinking about a range of future possibilities that humans usually don't think about and, in fact, assiduously avoid thinking about. We have explored some of the reasons why humans avoid engaging with the complexities and uncertainties of possible futures. We have concluded that much of the explanation lies in psychological defence mechanisms that lead us to imagine the world as being simpler and more predictable than it is and to filter the information provided by our senses so that only that which fits our simplified mental models reaches our consciousness.

Why should this matter? So what if we live in blissful ignorance? If we consider practice as not only doing, but also being, becoming and knowing, we will miss many opportunities for deeper and more meaningful contributions to society as professionals if we are not open to the ever-increasing possibilities that many plausible futures look like providing. Practice could evolve into very different things in different futures. To explore the range of possibilities, the authors of this book must themselves acknowledge and address their thinking constraints, but readers, including existing and future professionals, must also open their minds or risk missing the early signs of emerging opportunities and risks that, in some credibly envisaged futures, could be the difference between those who get to choose their career path and those who have it chosen for them or don't work at all.

Beyond employment prospects, encouraging unconstrained thinking about future relationships between knowledge, expertise and society is likely to be important for addressing social inequalities and inequities, many of which arise due to failure to imagine better possibilities and adherence to outdated attitudes and governance arrangements that encourage a struggle for power and influence, including the power than comes from restricting access to information.

All of the above pose the question: What sorts of futures for practice and its roles in society might be possible if the limitations of human thinking were overcome or at least alleviated? We have not tried to answer this question, but rather have explored how those limitations might be addressed. Other chapters in this book will consider what alternative futures might be like – but all authors will have struggled to some extent with the challenges identified in this chapter.

NOTES

[1] www.gsg.org
[2] Earthsea, also known as The Earthsea Cycle, is a series of fantasy books written by the American writer Ursula K. Le Guin: http://www.ursulakleguin.com/Index-Earthsea.html

REFERENCES

Althusser, L. (1971). *Lenin and philosophy and other essays* (B. Brewster, Trans.). London, England: New Left Books.

Ariely, D. (2008). *Predictably irrational*. London, England: Harper Collins.
Baird, J. (2018, November 9). Women are burning with a kind of cold fury. *The Sydney Morning Herald*. Retrieved from https://www.smh.com.au/politics/nsw/women-are-burning-with-a-kind-of-cold-fury-20181109-p50ezj.html
Biesta, G. J. J., Osberg, D., & Cilliers, P. (2008). From representation to emergence: Complexity's challenge to the epistemology of schooling. *Educational Philosophy and Theory, 40*(1), 213-227.
Chambers, I., Kubiszewski, I., Kenny, D. C., Maung, A. C., Costanza, R., Sofiullah, A., Hernandez, M., Yuan, K., Harte, S., Cork, S., Liao, Y., Finnigan, D., Htwe, T., Zingus, L., Kasser, T., & Atkins, P. (2018, in press). A public opinion survey of four future scenarios for Australia in 2050. *Futures*.
Cork, S., Grigg, N., Alford, K., Finnigan, J., Fulton, B., & Raupach, M. (2015). *Australia 2050: Structuring conversations about our future*. Canberra, Australia: Australian Academy of Science. Retrieved from https://www.science.org.au/files/userfiles/support/reports-and-plans/2015/australia-2050-vol-3.pdf
Costanza, R., Kubiszewski, I., Cork, S., Atkins, P. W. B., Bean, A., Diamond, A., Grigg, N., Korb, E., Logg-Scarvell, J., Navis, R., & Patrick, K. (2015). Scenarios for Australia in 2050: A synthesis and proposed survey. *Journal of Futures Studies, 19*(3), 49-76.
Czechmihaley, M. (2008). *Flow: The psychology of optimal experience*. New York, NY: Harper Perennial Modern Classics.
Eckersley, R. (2008). Nihilism, fundamentalism, or activism: Three responses to fears of the apocalypse. *The Futurist, January-February*, 35-39.
Edis, T. (2015, May 12). Why UWA was right to reject the $4m Lomborg bribe. *Climate Spectator*. Retrieved from http://www.businessspectator.com.au/article/2015/5/12/policy-politics/why-uwa-was-right-reject-4m-lomborg-bribe
Gilbert, D. (2010). *Harvard thinks big 2010 – Daniel Gilbert – global warming and psychology*. Retrieved from https://vimeo.com/10324258
Haraway, D. (1991). A cyborg manifesto: Science, technology, and socialist-feminism in the late twentieth century. In D. Haraway (Ed.), *Simians, cyborgs, and women: The reinvention of nature* (pp. 149-182). New York, NY: Taylor & Francis.
Hunt, D. V. L., Lombardi, D. R., Atkinson, S., Barber, A. R. G., Barnes, M., Boyko, C. T., Brown, J., Bryson, J., Butler, D., Caputo, S., Caserio, M., Coles, R., Cooper, R. F. D., Farmani, R., Taterell, M. R., Hale, J., Hales, C., Hewitt, C. N., Jankovic, L., Jefferson, I., Leach, J., MacKenzie, A. R., Memon, F. A., Sadler, J. P., Weingaertner, C., Whyatt, J., Duncan, R., & Christopher, D. F. (2012). Scenario archetypes: Converging rather than diverging themes. *Sustainability, 4*(4), 740-772.
Jones, N. A., Ross, H., Lynam, T., Perez, P., & Leitch, A. (2011). Mental models: An interdisciplinary synthesis of theory and methods. *Ecology and Society, 16*(1), Art. 46.
Le Guinn, U. (2001) *Always coming home*. Oakland, CA: University of California Press.
Macy, J. (2018). *Joanna Macy and the Great Turning*. Retrieved from http://www.joannamacyfilm.org/
Macy, J. R., & Johnstone, C. (2012). *Active hope: How to face the mess we're in without going crazy*. Sydney, Australia: Finch.
Mies, M., & Shiva, V. (1993). *Ecofeminism*. Melbourne, Australia: Spinifex.
Osberg, D. (2009). 'Enlarging the space of the possible' around what it means to educate and be educated. *Complicity: An International Journal of Complexity and Education, 6*(1), x00-00.
Pascale, R., Sternin, J., & Sternin, M. (2010). *The power of positive deviance: How unlikely innovators solve the world's toughest problems*. New York, NY. Harvard Business Review Press.
Peterson, G., Cumming, G., & Carpenter, S. R. (2003). Scenario planning: A tool for conservation in an uncertain world. *Conservation Biology, 17*(2), 358-366.
Raskin, P., Banuri, T., Gallopín, G., Gutman, P., Hammond, A., Kates, R., & Swart, R. (2002). *Great transition: The promise and lure of the times ahead*. Boston, MA: Stockholm Environment Institute.
Roxburgh, C. (2003). Hidden flaws in strategy. *The McKinsey Quarterly, 2003*(2), 27-39.
Royal Dutch Shell. (1995). Global scenarios 1995–2020. Retrieved from https://www.shell.com/energy-and-innovation/the-energy-future/scenarios/new-lenses-on-the-future/earlier-scenarios/_jcr_content/par/expandablelist/expandablesection_225706646.stream/1447230859664/8d

f8039be89c62fcc51f2d8ed3b224d1ab7f1db544464a31d13b47a537fd804b/shell-global-scenarios19952020.pdf

Scearce, D., Fulton, K., & Global Business Network Community. (2004). *What if? The art of scenario thinking for non-profits*. Emeryville, CA: Global Business Network. Retrieved from https://community-wealth.org/sites/clone.community-wealth.org/files/downloads/report-scearce-et-al.pdf

Schwartz, P. (2003). *Inevitable surprises: Thinking ahead in a time of turbulence*. New York, NY: Gotham.

Senge, P. M. (2010). *The Fifth Discipline: The art and practice of the learning organization*. New York, NY: Cornerstone Digital.

Shiva, V. (1993). *Monocultures of the mind: Perspectives on biodiversity and biotechnology*. London, England: Zed Books.

Snowden, D. (2002). Complex acts of knowing: Paradox and descriptive self-awareness. *Journal of Knowledge Management, 6*(2), 100-111.

Susskind, R., & Susskind, D. (2015). *The future of the professions: How technology will transform the work of human experts*. Oxford, England: Oxford University Press.

Taleb, N. N. (2007). *The black swan: The impact of the highly improbable*. Camberwell, Australia: Allen Lane.

Weaver, K., Garcia, S. M., Schwarz, N., & Miller, D. T. (2007). Inferring the popularity of an opinion from its familiarity: A repetitive voice can sound like a chorus. *Journal of Personality and Social Psychology, 92*(5), 821-833.

Wertheim, M. (1997). *Pythagoras's trousers: God, physics, and the gender war*. New York, NY: W.W. Norton & Company.

Zurek, M. B., & Henrichs, T. (2007). Linking scenarios across geographical scales in international environmental assessments. *Technological Forecasting and Social Change, 74*(8), 1282-1295.

Steven Cork PhD (ORCID: https://orcid.org/0000-0002-3270-4585)
Crawford School of Public Policy
Australian National University, Australia
Ecoinsights, Australia
Australia21, Australia

Debbie Horsfall PhD (ORCID: https://orcid.org/0000-0002-9266-6234)
Western Sydney University, Australia

STEVEN CORK AND KRISTIN ALFORD

3. PLAUSIBLE PRACTICE FUTURES

Hitherto [1848] it is questionable if all the mechanical inventions yet made have lightened the day's toil of any human being. They have enabled a greater population to live the same life of drudgery and imprisonment, and an increased number of manufacturers and others to make fortunes. They have increased the comforts of the middle classes. But they have not yet begun to effect those great changes in human destiny, which it is in their nature and in their futurity to accomplish. (John Stuart Mill, cited in Susskind & Susskind, 2015, p. xi)

At the time of the first Industrial Revolution, questions were being asked about the role of machines in *relieving* people from the burdens of work. As we enter what has been called a Fourth Industrial Revolution, defined by the rise of artificial intelligence (AI), concerns are being expressed about machines *taking* work from people. Work has assumed a larger role in our lives than ever before in human history (Beckett, 2018). Many commentators are asking what work might be like in the future, assuming that work will continue to play its central role in our lives. In comparison, others contend that we should be asking how future societies might best achieve what work currently achieves, and not necessarily assume that *human toil* will be an important part of these societies (Susskind & Susskind, 2015). In this book about *practice* the latter question prompts us to ask what social needs practice currently meets and how those needs might best be met in a range of plausible futures, including ones in which professional practice has little or no role.

In this chapter, we review a range of literature about the future of work, but focus on *practice* within *professions* as a particular aspect of work. If we consider the professions as bodies that have been established to help society deal with complex issues in a way that ensures relevant knowledge is generated and applied for the good of society (Susskind & Susskind, 2015), then we can think of *professional practice* as what members of *professions* do and how they do it. We begin by explaining how the discipline of *futures thinking* can help us consider multiple, plausible practice futures. We then structure the chapter around a generic process of futures thinking. We conclude that powerful drivers of change, including the unprecedented rise of both task oriented and cognitive digital technologies, are set to change all aspects of practice but that the possibilities are wider and more complicated than considered in the majority of the literature.

FUTURES THINKING

Futures thinking – thinking about the future in a structured and systematic way – is an emerging discipline, drawing its theory from many other disciplines (Cork,

2015; Hines & Gold, 2013; Ramos, 2004). Two important assumptions of futures thinking are that the future is unknowable in detail; and preparing for the future using prediction, as enticing as it is to humans (Taleb, 2007), is not only invalid but very risky over medium to long-term time horizons (P. Schwartz, 2003; Tetlock, 2005; van der Heijden, 2005). The discipline has been given various other names, including *Strategic Foresight*, *Scenario Planning,* and *Scenario Thinking*. In Chapter 2 of this book, Cork and Horsfall review some theory and practice within this and related disciplines.

Figure 3.1 shows a generic approach to futures thinking. It involves clarifying the questions to be asked about the future, considering potential drivers of change and the uncertainty surrounding them, and using structured narratives (scenarios) to explore multiple plausible futures and their strategic implications. The rest of this chapter is structured around the process shown in Figure 3.1.

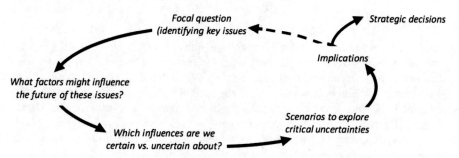

Figure 3.1. A generic process for futures thinking (adapted from Scearce, Fulton, & Global Business Network Community, 2004.)

QUESTIONS ABOUT THE FUTURE OF PRACTICE AND PROFESSIONS

As foreshadowed in our introduction, the key question guiding this chapter is: *How might future societies' needs for deep knowledge and practical expertise be met in the next 30 years and beyond?* To address this question, we will first review literature focused on the future of work as a central part of human life, but then broaden our focus to consider alternative futures in which work and even professional practice might be less pivotal. We follow Higgs (2012) in our interpretation of *practice*:

> For doctors, engineers, historians, priests, physicists, musicians, carpenters and many other occupational groups, practice refers to the activities, models, norms, language, discourse, ways of knowing and thinking, technical capacities, knowledge, identities, philosophies and other sociocultural practices that collectively comprise their particular occupation. (p. 3)

We pay particular attention to the concept of *professions*, which are currently regarded by many as the ultimate solution to help societies manage and apply complex knowledge. Professionals, such as doctors, lawyers, teachers, accountants, tax advisers, management consultants, architects, journalists and the clergy (Susskind & Susskind, 2015), have characteristics that set them apart from other people working in other occupations, including: having central ideas and mechanisms; regulatory autonomy granted by the state; shared knowledge and training; shared professional ethics and purpose; and status, power and identity (Rogers et al., 2016). Many authors discuss the "bargain" that society has struck with the professions that provide recognition and status in return for making sense of the complexity of many challenges facing society and helping to address these challenges for the common good. But it has been suggested that major challenges to professionalism and the professional class are emerging (Muzio, Ackroyd, & Chanlat, 2008; Rogers et al., 2016; Susskind & Susskind, 2015) and that we might already be seeing a process of *de-professionalisation* of these occupations as a social trend (Haug, 1972; Randall & Kindiak, 2008).

POTENTIAL DRIVERS OF CHANGE

In Table 3.1, we summarise several leading studies on factors thought likely to drive change in work and workplaces over a range of future time horizons.

Table 3.1. Drivers of change identified in seven recent analyses.

Driver/Trend	Publication[a] 1 2 3 4 5 6 7
Technologies: capabilities; convergence; connectivity; automation	X X X X X X X
Demography: ageing, longevity; diversity; migration; urbanisation	X X X X X X
Changing social values and worker expectations	X
Customer empowerment	X
Income uncertainty/inequality; shifts in economic power	X X X
New economies: rise of Africa; Asia demanding more services	X X X X
Changing economic perspectives vs. consumerism	X X X
Changing employment markets; globalisation of work and workers	X X X X
Growth of services industries	X
Personalisation of products/demand for crafts	X
Changing organisational structures/ecosystems	X X X X
Changing roles of political action	X
Resource scarcity, climate change, environment	X X X

[a] Sources (followed by time horizon of study): 1. PricewaterhouseCoopers (2017) [2030]; 2. Hagel, Schwartz, & Bersin (2017) [not stated]; 3. J. Schwartz (2016) [2020/30]; 4. Hajkowicz et al. (2016) [2036]; 5. Störmer, Rhisiart, & Glover (2014) [2030]; 6. Pearson et al. (2016) [2030s]; 7. Davies, Fidler, & Gorgis (2011) [2020].

Changing values and expectations loom large among these drivers. Workforces in developed countries over the next few decades are expected to have greater diversity of attitudes, experiences and expectations than any in the past, as four generations (Baby-Boomers and Generations X, Y and Z) work side by side (Störmer et al., 2014). Many historians, anthropologists and economists suggest that, especially in developed countries, *belief in work is crumbling among people in their 20s and 30s* and *more and more work feels pointless or even socially damaging* (Beckett, 2018). For most of human history, work was secondary to other aspects of life, but the modern work ethic makes it central. Ideas about redefining or eliminating work arose in the 1960s and 70s and have remerged since the early 2010s (including in Australia). Some see potential for life with much less human work, or no work at all, to be: "calmer, more equal, more communal, more pleasurable, more thoughtful, more politically engaged, more fulfilled" (Beckett, 2018, p. 3). Those imagining a post-work future look to policies such as a universal basic income to balance the effects of automation.

The potential for some/many jobs to be replaced by advanced mechanical and cognitive technologies is a hotly debated topic in the literature. One side of this debate forecasts massive unemployment as jobs for humans are taken by machines, while the opposite view is that this Fourth Industrial Revolution will, like those before it, create *new* jobs for humans to replace those lost (Autor, 2015; Schatsky & Schwartz, 2015). Frey and Osborne (2013), for example, estimated that 47% of jobs in the United States economy were vulnerable to computerisation. On the other hand, David Fagan concluded that job losses are likely to be as low as 20% when account is taken of new jobs that could be created by emerging challenges, such as mitigation of climate change, and retooling cities to cope with urbanisation and care needs of ageing Western societies (Fagan, 2017).

A good example is the crafting of surfboards (Warren & Gibson, 2018). In Australia, Hawaii and California, design and manufacture of surfboards has a major cultural component and requires the integration of culture with understanding of particular wave types at different locations. While this complexity might one day be within the capabilities of machines, for the foreseeable future the art of the human surfboard designer is likely to be important. We see this being played out as competition between large corporate surfboard manufactures and local designers/ manufacturers, with each serving different markets.

The distinction has been made in various literature between tasks and jobs – while tasks might become automated, an overall job itself might continue to require human oversight and/or might emerge as a different job performed by the same people. In Australia, it has been suggested that new jobs will outnumber job losses by 10 to 1 (Pearson et al., 2016). Frey and Osborne (2013) concluded that jobs involving the following are likely to resist computerisation: perception and manipulation (e.g. requiring manual dexterity); creative intelligence; and social intelligence (e.g. requiring social perceptiveness, persuasive and/or caring skills). The possible re-emergence of demand for artisanal skills (Beckett, 2018; Foundation for Young Australians, 2017a) could give humans an edge over computers. An occupation's probability of computerisation has been found to negatively relate to wages and educational attainment (Frey & Osborne, 2013). It is estimated that around 70% of young Australians are getting their first job in roles that will change fundamentally or disappear in the next 10–15 years due to automation, and that nearly 60% of Australian students are currently studying for occupations where at least two thirds of jobs will be automated (Foundation for

Young Australians, 2015). Most authors agree that unequal distribution of opportunities and rewards from technology will create major future challenges and that countries like Australia are poorly prepared to deal with them (Brighton-Hall & Hooper, 2017; Foundation for Young Australians, 2017b; Vardi, 2017).

Störmer et al. (2014) particularly identified potent *disruptors* that could radically change work, some of which are already exerting influence. These include: reverse migration (e.g. immigrants leaving industrialised nations); a set-change in employees' values (e.g. selecting employers based on values); zero-hour contracts becoming the norm; anywhere, anytime skills delivery; extreme effects of artificial intelligence (e.g. replacing surgeons and lawyers); de-globalisation (e.g. increased protectionist and nationalist tendencies); centres of excellence moving to emerging countries; disrupted Internet developments (e.g. greatly enhanced cybercrime); resource conflicts or climate disasters that threaten supply; and partial fragmentation of major economies (e.g. the European Union).

SCENARIOS FOR THE FUTURE OF WORK

Our awareness of possible futures can be expanded by considering how social, technological, economic, environmental, political, legal and other changes might interact with one another. One way to do this is to focus on so-called *critical uncertainties* – processes whose direction of development is uncertain but which are thought likely to have strong influences on many other drivers of change. Table 3.2 provides three key studies exploring futures of work in this way. These scenarios suggest that different attitudes of society and strength of communities' voices could make a big difference to how policies are developed and applied and how corporations treat their employees and society.

In some scenarios, it is the responsibility of individuals to train themselves and maintain their skills, while in other scenarios corporations and/or governments take this role. The current trend towards the gig economy is an example of how the former might look (Flanagan, 2017). In turn, the outcomes from corporations investing in training and retaining their staff are different depending on different economic and social settings. The scenarios give insights into how these different circumstances might arise and suggest early warning signs of different futures.

The scenarios also explore the possibilities of inequalities due to differences in opportunities or abilities to receive training and enter the workforce. Inequality is likely to be higher and harder to deal with, for example, in individualist worlds than collectivist worlds in the PricewaterhouseCoopers (PwC) scenarios. An unequal world is the clear outcome of the *Great Divide* scenario. The Data61 scenarios (Hajkowicz et al., 2016) raise the question of how society deals with displaced workers in the transition to new work futures. Other studies have explored the possible effects of occupational uncertainty on mental and physical health (Lewchuk, 2017).

Table 3.2. Three scenario sets exploring futures of work.

(Scenarios are developed by exploring combinations of the uncertainties – descriptions edited from originals.)

The Future of Work: Jobs and Skills in 2030 (United Kingdom)[a]	Tomorrow's Digitally Enabled Workforce (Australia)[b]	Workforce of the Future: Competing Forces Shaping 2030 (Global)[c]
Uncertainties	**Uncertainties**	**Uncertainties**
Economy, Labour market, Social conditions, Employment policies, Innovation uptake, Education and training	Institutional change (limited [1] vs. significant [2]) Task automation (low [A] vs. high [B])	Business fragmentation [1] vs. corporate integration [2] Individualism [A] vs. collectivism [B]
Scenarios	**Scenarios**	**Scenarios**
Forced Flexibility (business as usual): Greater business flexibility and incremental innovation lead to modest economic growth but less opportunities and weakened job security for low-skilled.	Lakes [1, A]: Linear advances in technology but uneven penetration. Little change to business structures or processes. Australian workforce similar to today.	Innovation Rules [1, A]: Innovation outpaces regulation. Race to satisfy consumers. Digital platforms give influence to winning ideas. Specialists and niche profit-makers flourish.
The Great Divide: Despite robust growth driven by high-tech industries, a two-tiered, unequal society has emerged.	Harbours [1, B]: Promises of AI and automation are fully realised. Technology has replaced many jobs, but there are fewer changes to employment models.	Corporate is King [2, A]: Big company capitalism as organisations grow and individual preferences trump social responsibility.
Skills Activism: Technological innovation and automation of professional work lead to large-scale job losses, prompting an extensive government-led skills programme.	Rivers [2, A]: Technology has advanced slower than envisaged. Task automation has had low impact on most jobs. But substantial change in organisational structures, cultures and practices.	Companies Care [2, B]: Business driven by social responsibility, trust and concern about demography, climate and sustainability. Humans Come First [1, B]: Social-first and community businesses prosper. Crowd-funding of ethical brands. A search for meaning and relevance with a social heart. Artisans and "new Worker Guilds" thrive. Humanness is valued.
Innovation Adaptation: In a stagnant economy, productivity is improved through a systematic implementation of ICT solutions.	Oceans [2, B]: Exponential growth in technology. Innovative, socially inclusive employment. Amazing opportunities for individuals and society.	

[a] Störmer et al. (2014) [b] Hajkowicz et al. (2016) [c] PricewaterhouseCoopers (2017)

Three of the scenarios in Table 3.2, *Lakes*, *Rivers* and *Humans Come First*, envisage futures in which the expected rise of AI is slower than expected, has uneven penetration or is reversed. We suggest in our conclusion that, as improbable as these scenarios appear at present, failure to rehearse their implications could leave society vulnerable to highly undesirable disruptive surprises. Even if extreme forms of these scenarios don't eventuate, there is much to be learned from considering divergences from the generally accepted future. And we should not overlook the potential emergence of *transition* jobs – jobs that emerge to help people move from one future to another.

FUTURES OF PROFESSIONS

For several decades, authors have argued that *de-professionalisation* of the professions has been occurring, driven largely by societies' increasing demands for direct access to information and expertise combined with electronic media making that access more possible (Haug, 1972; Randall & Kindiak, 2008). Susskind and Susskind (2015) speculate about where this process might lead in short, medium and longer time frames. Table 3.3 has been created here to illustrate these ideas. They see three main factors – the market, technology and human ingenuity – driving an evolutionary process, in which ownership and control of knowledge and expertise move from craft-like, exclusive professions to the whole of society.

Table 3.3. Possible evolution of professional work.

Stage	When?	Characteristics
Professions as crafts	Past	Expert, bespoke services; knowledge and its application controlled
Standardisation	Emerging now	Sharing of knowledge; agreement on approaches; standardised checklists
Systematisation	Early signs now	Systems, technology and tools assist, or replace, human experts in carrying out professional tasks
Externalisation	Medium to long-term future	Online services, charged for Online services for no charge Free access to information commons

Already, Susskind and Susskind (2015) argue, boundaries between and within professions and between humans and machines are breaking down (e.g. paramedics, medical lawyers, robotic pharmacists). They expect that knowledge increasingly will be generated and shared across groups of people with diverse skills and experience, often working as multidisciplinary collaborations. Eventually, in their view, professions won't be needed to help society understand complex bodies of data and much practical expertise will be available, in a variety of ways, online. Ultimately, they anticipate, knowledge will be not only managed, but also generated, by machines, because this will be found to be the most effective and efficient way to meet society's needs.

The evolution described in Table 3.3 could take several decades to unfold, and could stall or accelerate at different times and in different places and different professions. Furthermore, rather than being a single pathway, there are many ways in which the evolution could unfold. Susskind and Susskind (2015) have suggested seven models of how knowledge might be generated, managed and applied. Figure 3.2 builds on the ideas of these authors and others (see below). When, where and how these models might emerge are likely to be influenced not only by technology but also by multiple drivers of change.

In Figure 3.2 we illustrate how Susskind and Susskind's ideas can be combined with scenario-thinking to challenge and broaden conceptions about how the future *might* unfold. We have included the critical uncertainties used in PwC's scenarios (i.e. as shown in Table 3.2). These critical uncertainties have been used often in recent scenario projects because they are considered likely to have important influences on how society might generate and/or respond to major challenges like climate change, population growth, designing liveable future cities, sustainability and inequality (Costanza et al., 2015; Intergovernmental Panel on Climate Change, 2000; Office for Science and Technology, 2002).

In positioning the models in Figure 3.2, we assume, for example, that the traditional model of the professions might be favoured in *individualist-fragmented* futures if professions are considered necessary to protect society from unscrupulous practices or to maintain human empathy against market forces. Models requiring networking might require *connected governance* across society and might be inhibited by *fragmented governance* arrangements. The *Knowledge Engineering* and *Communities of Experience* models would require trust and cooperation, which might be harder to generate in highly *individualistic* futures.

As Susskind and Susskind (2015) acknowledged, evolution from one model to another depends largely on incentives for producing new knowledge and expertise if those generating it cannot have exclusive ownership, and whether potential investors can afford to fund generation of new knowledge and expertise. The answers to these questions will differ in different parts of Figure 3.2 and in other combinations of future uncertainties like those considered in Tables 3.1 and 3.2. Figure 3.2 is just one example of how considering future uncertainties might help us consider alternative ways in which futures might unfold, rather than taking the comfortable and common, but flawed, approach of predicting the most likely future.

CONCLUSION

Most authors agree that many aspects of what humans currently do in their employment will, eventually, be replaced by machines and AI. The major differences of opinion are around how quickly this might happen; how complete the replacement might be; what sorts of new roles might appear for humans; and how human values such as ethics, morality, empathy and compassion might be exercised in future societies. The most optimistic view in the current commentaries is that we will see new relationships between humans and machines that produce better outcomes for humans. As with previous industrial revolutions, there is likely to be some social disruption as this better world emerges.

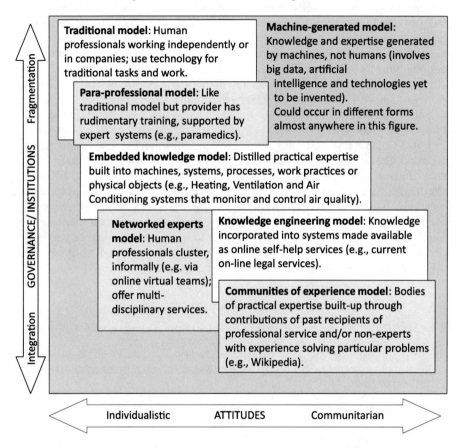

Figure 3.2. How seven models of future knowledge generation and application might be favoured under different combinations of societal attitudes and governance.

Other chapters in this book (e.g. Chapters 5 and 15) discuss the short- to medium-term challenges for different generations who will live through the next few decades of transition. The challenges are likely to be different for different generations and social cohorts, but two key issues emerge: (1) the skills required for employment in future workplaces are likely to be increasingly multidisciplinary and related more to problem solving and less to task-oriented occupations; and (2) such jobs might employ only a small proportion of the population, meaning that society will need to be restructured to avoid inequality of opportunity to lead fulfilling lives.

These challenges will require that societies think carefully about such issues as: the need for revolutionary changes to higher education and other training institutions to allow education across disciplinary boundaries and teaching that is not limited to fixed locations; rethinking about the extent to which governments or industry should take responsibility for training; matching education with real-world needs and expectations; marrying academic training with opportunities to apply new knowledge and skills; supporting lifelong learning and rapid retraining, and opportunities for self-directed learning; and new credentialing systems to allow for all of the above (Enders & Musselin, 2008; PwC Australia, 2016; Rainie & Anderson, 2017; J. Schwartz, 2016; Stephan, Kamen, & Bannister, 2017).

Society should be careful to not accept, uncritically, the future believed to be most likely – that jobs and professions will be progressively replaced by machines and AI. Some version of this future looks probable, but major uncertainties surround the nature, timing and location of the necessary transitions. These uncertainties should be considered and the plausible alternative pathways to alternative futures rehearsed so that surprises are reduced and preparedness increased. At least some futures-thinking reviewed here (e.g. the *Lakes*, *Rivers* and *Humans Come First* scenarios in Table 3.2) suggests that, while surprises are most likely to come as new technologies and ways to use them, we could be caught off-guard if we fail to be watching for futures in which the expected AI-driven world does not emerge or emerges more slowly than expected.

REFERENCES

Autor, D. H. (2015). Why are there still so many jobs? The history and future of workplace automation. *Journal of Economic Perspectives*, 29(3), 3-30.

Beckett, A. (2018, January 19). Post-work: The radical idea of a world without jobs. *The Guardian*. Retrieved from https://www.theguardian.com/news/2018/jan/19/post-work-the-radical-idea-of-a-world-without-jobs

Brighton-Hall, R., & Hooper, N. (2017, August 31). Australians are not ready for the future of work. *Australian Financial Review*. Retrieved from https://www.afr.com/brand/boss/we-are-not-ready-for-the-future-of-work-20170809-gxswin

Cork, S. (2015). Using futures thinking to support ecosystem assessments. In M. Potschin, R. Haines-Young, R. Fish, & R. K. Turner (Eds.), *Routledge handbook of ecosystem services* (pp. 170-187). London, England: Routledge.

Costanza, R., Kubiszewski, I., Cork, S., Atkins, P. W. B., Bean, A., Diamond, A., Grigg, N., Korb, E., Logg-Scarvell, J., Navis, R., & Patrick, K. (2015). Scenarios for Australia in 2050: A synthesis and proposed survey. *Journal of Futures Studies*, 19(3), 49-76.

Davies, A., Fidler, D., & Gorgis, M. (2011). *Future work skills 2020*. Palo Alto, CA: Institute for the Future. Retrieved from http://www.iftf.org/uploads/media/SR-1382A_UPRI_future_work_skills_sm.pdf

Enders, J., & Musselin, C. (2008). Back to the future? The academic professions in the 21st century. In OECD (Ed.), *Higher education to 2030, Volume 1, Demography* (pp. 125-150). Paris, France: OECD Publishing.

Fagan, D. (2017, November 1). Will technology take your job? New analysis says more of us are safer than we thought, but not all. *The Conversation*. Retrieved from http://theconversation.com/will-technology-take-your-job-new-analysis-says-more-of-us-are-safer-than-we-thought-but-not-all-86219

Flanagan, F. (2017). Symposium on work in the 'gig' economy: Introduction. *The Economic and Labour Relations Review, 28*(3), 378-381.
Foundation for Young Australians. (2015). *The new work order: Ensuring young Australians have skills and experience for the jobs of the future, not the past.* Retrieved from https://www.fya.org.au/report/new-work-order/
Foundation for Young Australians. (2017a). *The new basics: Big data reveals the skills young people need for the new work order.* Retrieved from https://www.fya.org.au/wp-content/uploads/2016/04/The-New-Basics_Update_Web.pdf
Foundation for Young Australians. (2017b). *The new work smarts: Thriving in the new work order.* Retrieved from https://www.fya.org.au/report/the-new-work-smarts/
Frey, C. B., & Osborne, M. A. (2013). The future of employment: How susceptible are jobs to computerisation? *Technological Forecasting and Social Change, 114C*, 254-280.
Hagel, J., Schwartz, J., & Bersin, J. (2017). Navigating the future of work. *Deloitte Review, 21*, 26-40.
Hajkowicz, S. A., Reeson, A., Rudd, L., Bratanova, A., Hodgers, L., Mason, C., & Boughen, N. (2016). *Tomorrows digitally enabled workforce.* Brisbane, Australia: CSIRO.
Haug, M. R. (1972). Deprofessionalization: An alternate hypothesis for the future. *The Sociological Review, 20*(S1), 195-211.
Higgs, J. (2012). Practice-based education: The practice-education-context-quality nexus. In J. Higgs, R. Barnett, S. Billett, M. Hutchings, & F. Trede (Eds.), *Practice-based education: Perspectives and strategies* (pp. 3-12). Rotterdam, The Netherlands: Sense Publishers.
Hines, A., & Gold, J. (2013). Professionalizing foresight: Why do it, where it stands, and what needs to be done. *Journal of Futures Studies, 17*(4), 35-54.
Intergovernmental Panel on Climate Change. (2000). *Emissions scenarios. Summary for policy makers.* Cambridge, England: Cambridge University Press.
Lewchuk, W. (2017). Precarious jobs: Where are they, and how do they affect well-being? *The Economic and Labour Relations Review, 28*(3), 402-419.
Muzio, D., Ackroyd, S., & Chanlat, J.-F. (2008). Introduction: Lawyers, doctors and business consultants. In D. Muzio, S. Ackroyd, & J-F. Chanlat (Eds.), *Redirections in the study of expert labour* (pp. 1-28). London, England: Palgrave Macmillan.
Office for Science and Technology. (2002). *Foresight futures 2020.* London, England: Department of Trade and Industry.
Pearson, L., Vonthethoff, B., Rennie, M., & Nguyen, T. (2016). *The future of work.* Canberra, Australia: Regional Australia Institute and NBN Australia's Broadband Network. Retrieved from http://www.regionalaustralia.org.au/home/the-future-of-work/
PricewaterhouseCoopers Australia (PwC Australia). (2016). *Australian higher education workforce of the future.* Retrieved from https://www.aheia.edu.au/cms_uploads/docs/aheia-higher-education-workforce-of-the-future-report.pdf
PricewaterhouseCoopers (PwC). (2017). *Workforce of the future – The competing forces shaping 2030.* England: PwC. Retrieved from https://www.pwc.com/people
Rainie, L., & Anderson, J. (2017). *Future of jobs and jobs training.* Retrieved from http://www.pewinternet.org/2017/05/03/the-future-of-jobs-and-jobs-training/
Ramos, J. M. (2004). *Foresight practice in Australia: A meta-scan of practitioners and organisations.* Melbourne, Australia: Australian Foresight Institute, Swinburne University.
Randall, G. E., & Kindiak, D. H. (2008). Deprofessionalization or postprofessionalization? Reflections on the state of social work as a profession. *Social Work in Health Care, 47*(4), 341-354.
Rogers, J., Kingsford-Smith, D., Clarke, T., & Chellew, J. (2016). *The promise of professionalism in the 21st century.* Working paper to be presented at the *Modern Professional Practice and Its Future* Conference. Retrieved from https://clmr.unsw.edu.au/sites/default/files/attached_files/rogers_and_kingsford_smith_et_al_the_promise_of_professionalism_-_summary_from_march_2016_symposium.pdf

Scearce, D., Fulton, K., & Global Business Network Community. (2004). *What if? The art of scenario thinking for non-profits*. Emeryville, CA: Global Business Network. Retrieved from https://community-wealth.org/sites/clone.community-wealth.org/files/downloads/report-scearce-et-al.pdf

Schatsky, D., & Schwartz, J. (2015). Redesigning work in an era of cognitive technologies. *Deloitte Review*, *17*. Retrieved from https://www2.deloitte.com/content/dam/insights/us/articles/work-redesign-and-cognitive-technology/DUP1203_DR17_RedesigningWorkCognitiveTechnologies.pdf

Schwartz, J. (2016). *The future of the workforce*. England: Deloitte. Retrieved from https://www2.deloitte.com/content/dam/Deloitte/global/Documents/HumanCapital/gx-hc-future-workforce.pdf

Schwartz, P. (2003). *Inevitable surprises: Thinking ahead in a time of turbulence*. New York, NY: Gotham.

Stephan, A., Kamen, M., & Bannister, C. (2017). Tech fluency. *Deloitte Review*, *21*, 18-93.

Störmer, E., Rhisiart, M., & Glover, P. (2014). *The future of work: Jobs and skills in 2030*. London, England: UK Commission for Employment and Skills. Retrieved from https://www.gov.uk/government/uploads/system/uploads/attachment_data/file/303334/er84-the-future-of-work-evidence-report.pdf

Susskind, R., & Susskind, D. (2015). *The future of the professions: How technology will transform the work of human experts*. Oxford, England: Oxford University Press.

Taleb, N. N. (2007). *The black swan: The impact of the highly improbable*. Camberwell, Australia: Allen Lane.

Tetlock, P. E. (2005). *Expert political judgement: How good is it? How can we know?* Princeton, NJ: Princeton University Press.

van der Heijden, K. (2005). *Scenarios: The art of strategic conversation*. Chichester, England: John Wiley & Sons.

Vardi, M. Y. (2017, September 2). What the Industrial Revolution really tells us about the future of automation and work. *The Conversation*. Retrieved from https://theconversation.com/what-the-industrial-revolution-really-tells-us-about-the-future-of-automation-and-work-82051

Warren, A., & Gibson, C. (2018). *Surfing places, surfboard makers: Craft, creativity and cultural heritage in Hawai'i, California and Australia*. Hawai'i: University of Hawai'i Press.

Steven Cork PhD (ORCID: https://orcid.org/0000-0002-3270-4585)
Crawford School of Public Policy
Australian National University, Australia
Ecoinsights, Australia
Australia21, Australia

Kristin Alford PhD
MOD., University of South Australia, Australia

PETER GOODYEAR AND LINA MARKAUSKAITE

4. THE IMPACT ON PRACTICE OF WICKED PROBLEMS AND UNPREDICTABLE FUTURES

Education faces a conundrum. On the one hand, imagined futures are becoming more diverse, fluid and contested. On the other, knowledge and learning are widely believed to be key to survival and success. Speculations about future patterns of employment vary considerably, with uncertainties about the breakdown between primary production, manufacturing and service work; paid and unpaid work; creative and routine work. Much is being written about the future of work itself and the future shape of working practices: about how work will be distributed between regions of the world, between rich and poor, between people and machines, about how work will be done and what it will mean for stakeholders. How then should education be reconfigured? There is a broad consensus that it cannot stay the same (Collins, 2017). But in many countries, there is deep disquiet about relations between current education and the futures of those it is meant to serve. Indeed, we can sense a paralysis, brought about by conflicting ideologies and by the intrinsic difficulties of making sense of an uncertain, complex world. Among many position statements that have been promulgated by governments, non-government organisations, business groups and others, we see key issues captured in the following:

> Young people need a *wider range of competences* than ever before to flourish, in a globalised economy and in increasingly diverse societies. Many will work in jobs that do not yet exist. Many will *need advanced linguistic, intercultural and entrepreneurial* capacities. *Technology* will continue to change the world in ways we cannot imagine. Challenges such as *climate change* will require radical adaptation. In this increasingly complex world, creativity and the ability to continue to learn and to innovate will count as much as, if not more than, specific areas of knowledge liable to become obsolete. *Lifelong learning* should be the norm. (European Commission, 2008, p. 3, emphasis added)

There are four significant points embedded in this statement. First, flourishing in the future will depend upon personal competencies. Second, each person's skillset will need continual updating: people should expect to spend time, and probably money, in regular rounds of lifelong learning. Third, so-called "soft skills" (which often turn out to be very hard to define, learn and assess), need at least as much attention as specific areas of expertise. People who cannot demonstrate these soft skills will probably be at a disadvantage. Fourth, complex changes whose precise implications cannot be predicted – technology, climate, demography – will require people to become more adaptable. Each of these concerns appears in many recent documents of this kind (Organisation for Economic Co-operation and Development [OECD], 2018). It is not surprising that people responsible for educational policy, and for exercising high-level leadership in areas such as curriculum and assessment, have been struggling with the implied challenges. A cynical observer might say that leadership has been abandoned: the problem outsourced to young people, to sort out their own paths through the future, just as problems of unsustainable growth, environmental damage, inequality and underinvestment in infrastructure have been deferred by the current ruling generations.

In this chapter, we aim to offer something more positive. We suggest that there are tools that people can learn to use to deal with complex "wicked" problems. These tools can be used by young and old, but are especially relevant to those who are invested in a problematic situation – those with "skin in the game". These ways of dealing with wicked problems are deeply *social*. They do not start from an assumption that the best problem solvers are lone wolves: creative, entrepreneurial market-disruptors, motivated by personal profit. Quite the reverse. In our view, tools for working on wicked problems are embodiments of shared "moral know-how", sharpened for the work of collaborative and cooperative future-making. The rapid and accelerating pace of technological development has had an odd effect on ways we imagine the future. We see it as unknowable and full of risks for which we should prepare, without really knowing what to prepare for: as if the explosion of technological possibilities creates a blinding glare. It need not be so. Technological profusion should cause us to ask a different kind of question: not "what will the future world be like, and require of us?" but "what kind of future world do we want to make?". The genre changes from prediction to design; from reading tea leaves to taking action. This needs tools and methods, which are learnable.

WICKED PROBLEMS

The term "wicked problem" is quite familiar in everyday discourse. However, it is one of those evocative phrases that has drifted away from its original moorings: wicked problems are not merely complex, difficult or persistent. The phrase originates in a paper by Rittel and Webber (1973) in the planning literature. Their goal was to distinguish a qualitatively different kind of problem, commonly faced in planning, social policy and other areas, and not properly reflected in contemporary descriptions of problem solving as a rational, scientific process. Rittel and Webber acknowledged that, in trying to succinctly distinguish "wicked" problems from "tame" problems, they drew some caricatures of prevailing "scientific" planning methods in order to provide sharper contrast.[1] That said, the characteristics listed in Table 4.1 accumulate to make a distinctive and recognisable class of problem: one that can rarely be fully resolved. There is a degree of overlap between some of these characteristics, but in sum we can say that wicked problems defy efforts to put boundaries around them and to identify their root causes; they sit within open causal/explanatory webs, and agreement on solutions varies between groups of people with different interests and values. The question then arises: how can students be helped to develop the capabilities needed to work, with others, on wicked problems? We approach this by considering the educational strategies that have been used to help students work with other kinds of problems.

Table 4.1. Ten characteristics of wicked problems (after Rittel & Webber, 1973).

Characteristic	Explanation
No definitive formulation	With "tame" problems, the problem solver can be given in advance all the information needed to solve the problem. With wicked problems, the information needed depends on candidate solutions. In other words, "the formulation of a wicked problem *is* the problem" (p. 161).
No "stopping rule"	No criteria inherent in the problem show unequivocally that it has been solved; work on wicked problems typically stops because of external constraints – such as time or money.
No "true-false" solutions	There are no formal decision rules that can be applied objectively to say that a solution is appropriate; different people/groups will have different views on a solution, but none will be able to finally determine its correctness.
No satisfactory tests of solutions	Any solution will generate waves of consequences; evaluation of these consequences will often cause people to reconsider what they previously saw as a satisfactory solution.
Every attempt at a solution has consequences	Every solution (or partial solution) has consequences – it changes the nature of the problem.
Inexhaustible solutions	Any new idea may become a candidate solution, or part of the solution; one cannot enumerate all possible solutions or solution steps in advance.
Every problem is essentially unique	Although the current problem may look like a previously experienced problem, there may always be a distinguishing property that turns out to be crucial. Premature "classification" of a problem is fraught with danger.
Every wicked problem can be a symptom of another problem	Removing discrepancies between an actual and a desired state of affairs usually involves identifying and removing causes; but removing apparent causes often creates a new problem – a new symptom of a deeper problem.
No definitive explanations of discrepancies	Many explanations for discrepancies are possible; each explanation favours a different solution; it is often impossible to be sure which explanation is best.
There is no right to be wrong	Science advances through the refutation of hypotheses; as long as scientists play by the rules of the game, they are not *blamed* for advancing a hypothesis which is subsequently refuted; people working on real-world wicked problems do not enjoy this immunity – they are held liable for the consequences of their plans and actions.

WICKED PROBLEMS, LEARNING AND WORK

Many aspects and examples of work – paid and unpaid – are taking on an epistemic quality: they involve the creation of new knowledge; sometimes in mundane ways, and sometimes with insights that are startling or profound (Markauskaite & Goodyear, 2017). Esko Kilpi (2016) puts it even more starkly:

> Post-industrial work is *learning*. Work is figuring out how to *define* and *solve* a particular problem and then scaling up the solution in a reflective and iterative way – with *technology* and alongside other *people*. (p. 34, emphasis added)

There are more capabilities buried in this assertion than may be immediately apparent. For one thing, "learning" in Kilpi's text is not just "learning from instruction". There is more to "figuring out" than typically occurs when learning by listening to someone else's explanation of settled facts. Neither does "learning to learn" do it justice. This often means little more than learning to manage oneself as a learner who does not have to rely on the supports and structures of educational institutions; in other words, learning to be an autonomous lifelong learner. This is valuable, but insufficient, for Kilpi is talking about learning to *frame* and *inquire*, learning to generalise solutions by co-creating shared technologies and practices. This involves creating new understandings – new knowledge – while working with other people and using appropriate epistemic tools and methods.

Acknowledging the existence, importance and widespread occurrence of *wicked* problems means going one step further in the list of capabilities required by contemporary learners (Markauskaite & Goodyear, 2017). It is not enough to be able to learn, to manage one's own learning and to adopt culturally approved methods of inquiry. Wicked problems, by their very nature, mean that people have to become more adept at *designing* inquiry. That is, working with others on wicked problems necessitates on-the-fly reconfiguration of methods and tools for inquiry. To be able to work together on wicked problems, people have to be able to make reasoned choices about the tools and methods they will jointly use to make progress on understanding the nature of the problem and candidate solutions.

It is important to understand that this has at least two kinds of implications for educational practice. The more obvious implications are concerned with curriculum design and what personal capabilities students might be helped to develop for themselves: knowing a variety of approaches to the design of inquiry, for example. But too strong a focus on personal skills and knowledge may distract attention from something deeper that is intrinsic to working on systemic wicked problems. Klaus Krippendorff (2006) discusses this in the course of his argument for a paradigm shift in the design professions. He links Rittel's account of wicked problems to the need for designers to move from first-order to second-order understanding: from understanding the surface properties of designed objects to understanding what designed things *mean* to the people who use them.

> Understanding someone else's understanding is an *understanding of understanding*, an understanding that recursively embeds another person's understanding in one's own, even if, and particularly when, those understandings disagree ... (Krippendorff, 2006, p. 66)

So the linked practices of *designing inquiry* and simultaneously *designing solutions* need to be conceived broadly enough to include people working together to achieve a degree of shared understanding and agreement on how they can go about solving problems and what they can collectively do – what they can propose, what they can make – in response to each problem. Learning to participate in such work is best understood in an *expansive* way: it should include knowing how to help identify, bring together and mobilise the people and things implicated in the problem and its candidate solutions: helping map and animate the socio-material networks involved.[2]

EDUCATIONAL STRATEGIES FOR LEARNING TO SOLVE PROBLEMS

Different kinds of knowledge need to be learned in different ways. One cannot learn to ski without skiing. (Someone who has never skied cannot reasonably say "I know how to ski".) One cannot learn historical inquiry simply by living through some of history. Different kinds of problems require different knowledge and different solution strategies. Some problems are amenable to solution using tried-and-tested methods. Others – including wicked problems – are not so easily dealt with. Consequently, students need to gain experience in dealing with a variety of kinds of problems. Through appropriate combinations of instruction, guidance and practice, they stand a chance of getting better at independently recognising what kind of problem they are confronting and what kinds of solution methods are likely to be appropriate. From an educational design point of view, it can be very helpful to have a repertoire of educational strategies for helping students learn how to recognise and make some progress with different kinds of problems. Problems can be classified in different ways, but the four-fold classification in Table 4.2 is enough to illustrate our general argument. The problem type is shown in bold. The matching educational strategy is in italics. The top row shows two types of classical problems; the bottom row shows two types of complex problems.

Well-structured Problems

In the classic literature on problem solving and learning to solve problems, a useful distinction is made between "well-structured" and "ill-structured" problems (Jonassen, 2011). School curricula offer plenty of examples of well-structured problem solving; indeed, this class of problem solving can be referred to as "recognition" or "transformation" problem solving. Once the student recognises the *type* of problem they have been given, they can apply a well-used set of problem-transformation moves to get to a solution. Solving simultaneous equations or generating proofs in geometry are cases in point. Educational strategies for helping students deal with well-structured problems include exposing them to multiple examples of each important problem type, providing guidance on how to recognise/classify each type of problem, and providing practice with the routine methods of solving each problem type.

In recent decades, educational technologies have proved quite effective at helping students develop the skills needed for working on well-structured problems. Automated computer-based marking of multiple choice questions and intelligent monitoring of students' problem-solving steps have been notable achievements in the application of artificial intelligence techniques in education (Micarelli, Stamper, & Panourgia, 2016). That said, the progress made in this area has benefited from the ease of representing routine problem-solving methods in algorithmic form. In other words, it is relatively easy to automate the teaching of skills which are themselves easy to automate in the workplace.

Table 4.2. Problem types and educational strategies (adapted from Markauskaite & Goodyear, 2017).

Well-structured problems	Ill-structured problems
Well-structured problems present all of the information needed to solve the problems in the problem representation; they require the application of a limited number of regular and circumscribed rules and principles that are organised in a predictive or prescriptive way; they possess correct, convergent answers; they have a preferred, prescribed solution process (Jonassen, 2011).	Conflicting goals, multiple solution methods, unanticipated problems, multiple forms of problem representation, one or more unknown problem elements, multiple solutions, uncertainty about knowledge applicable to the problem and a need for personal judgements or reliance on personal beliefs.
Recognition and transformation of problems	*Knowledge-design problems*
Tame problems	**Systemic wicked problems**
A tame problem is one where all the parties involved can agree what the problem is ahead of the analysis and which does not change during the analysis (see Ison, 2008, p. 146).	Wicked problems often involve defining the problem and solution simultaneously. They change during problem-solving work. Establishing agreement on the nature of the problem, appropriate methods for tackling it and criteria that can be used to know when a satisfactory solution has been reached are key aspects of working on wicked problems.
Epistemic framing problems	*Inquiry design problems*

Ill-structured Problems

Ill-structured problems are not as amenable to routine solution. They may be ill-structured for a number of reasons, including missing or unreliable information, or the availability of multiple solutions, such as when one wants to start a new business or redesign an existing service. Crucially, they cannot be solved simply by recognising the type of problem and invoking routines that are known to solve problems of that type. One might call them "knowledge design" problems: working out what additional knowledge might be relevant is an important capability here. Educational strategies for helping students take a designerly stance towards knowledge include providing scaffolding for various forms of collaborative knowledge-building and inquiry (see e.g. Scardamalia & Bereiter, 2014). Within higher education, enthusiasm for engaging undergraduate students in authentic research projects can be seen, in part, as providing opportunities for working on knowledge design (Brew, 2013).

Tame Problems

The bottom row of Table 4.2 distinguishes "tame" from "wicked" problems. An important quality of tame problems is that they are relatively stable: they do not change while being worked upon. They may be quite complicated, but they can be tackled with established methods of inquiry: specialised methods for researching and recommending solutions for the problem are available, such as are found at the core of many disciplinary and some professional practices. An example would be diagnosing a disease, in medical practice. Educational strategies for helping students master such capabilities include learning how to use the *epistemic frames* of a discipline or profession by gaining direct personal experience of playing its characteristic epistemic games. David Shaffer (2006) explains it thus:

> ... epistemic frames include methods for justification and explanation, and forms of representation, but orchestrated with strategies for identifying questions, gathering information, and evaluating results, as well as self-identification as a person who engages in such forms of thinking and ways of acting. If epistemic understanding and epistemic structures form the core of disciplines or subjects such as mathematics or history, then epistemic frames are the organizing principle for practices. Geometers, economists, statisticians, and engineers (all of whom use mathematics) have distinct epistemic frames that incorporate different epistemic understandings and structures from the domain of mathematics. (p. 228)

Systemic Wicked Problems

The bottom right cell of Table 4.2 brings us back to wicked problems and to a specific *take* on wicked problems that we call *systemic wicked problems*.

> Such problems are not uncommon in professional practice. Wicked systemic problems occur when development of an understanding of the problem, and enactment of candidate solutions, are distributed across multiple actors and across time. Wicked systemic problems change over time, sometimes in response to the (problem-solving) actions taken; therefore, a complete strategy cannot be worked out in advance. Joint action and shared enquiry (deep collective learning) are both needed – often in a loosely yoked way. (Goodyear & Markauskaite, 2018, p. 35)

Educational strategies that provide students with opportunities to work with systemic wicked problems include the use of Soft-Systems Methodology (SSM: see e.g. Checkland & Poulter, 2006), and in particular learning to collaboratively develop methods for inquiring into newly experienced wicked problems (Markauskaite & Goodyear, 2017). As Blackmore and Ison (2012) say, this is

> ... mainly about developing a critical appreciation of situations with others, recognizing what actions are systemically desirable and culturally feasible and getting organised to affect change in a positive way. (pp. 348-349)

In other words, because systemic wicked problems are usually distributed across, and solved by, many people working over extended periods of time, they necessarily involve both deep collective learning and joint action. While it is possible to provide students with artificially constructed problem-solving situations in which they can come to appreciate some collective learning and inquiry design methods (of SSM, for example), it is much harder to provide *authentic* opportunities to do such work. One cannot easily gain experience working as a stakeholder in a systemic wicked problem unless one is actually affected by the problem and candidate solutions.

If taken seriously, this has significant implications for curriculum design – and indeed for how we think about the framing of educational opportunities and the responsibilities of educational institutions. In short, one cannot gain genuine experience of working on systemic wicked problems unless one has some power to change the world.[3] Ison and others have made some progress on this by engaging students in authentic redesign of their own learning paths (Blackmore & Ison, 2012; Ison 2008). Authentic use of inquiry design approaches is possible in continuing professional development contexts, where practitioners as students can jointly investigate, and make changes in, areas of their professional work. By extension, one might think about ways of giving pre-practice students experiences of designing their own inquiry and learning paths while taking on authentic roles in their own professional learning communities.[4] For example, we have been applying SSM in our own teaching, aimed at helping students learn to lead educational innovation, by creating for them authentic opportunities to construct their own group work, inquiry processes, and joint learning and inquiry environment (Markauskaite, 2016).

LEARNING HOW TO CONSTRUCT PRODUCTIVE EPISTEMIC ENVIRONMENTS

Learning how to collaborate in the design of novel forms of inquiry is only partly a matter of personal skills and understanding. If work on the solution of wicked problems is understood as (a) socially distributed (e.g. involving a team of people) and (b) complex enough to require specialist tools (e.g. modelling, data analysis and visualisation packages), then students need experience of processes involved in bringing together the right mix of people and tools. In other words, learning to work on wicked problems also involves constructing the right environment for the work. Insofar as collaboratively designing and conducting inquiry features within work on wicked problems, we can say that students can benefit from learning how to construct or configure productive epistemic environments (Goodyear & Markausakaite, 2018). This is too broad an area for us to deal with comprehensively in the current chapter, so we will concentrate on *epistemic instruments*. We draw on the theory of instrumental genesis (Rabardel & Bourmand, 2003) to outline what is involved. The theory of instrumental genesis is an extension of activity theory, focused on the role of tools in mediated activity. On this view, most human activity is directed towards a concrete goal (an "object") but is accomplished indirectly through the use of mediators – notably through the use of tools. They introduced the useful analytic distinction between a tool (or, more generally, an artefact) and its utilisation schemes. An *artefact* – such as a hammer, X-ray machine or laptop computer – has a number of relatively fixed material (and/or digital) properties that may constrain and suggest its uses. A person develops ways of using artefacts. They may, for example, use a hammer for driving home screws or pulling out nails, in addition to its more common applications. These habitual ways of using each artefact are organised as *utilisation schemes*: personal methods for carrying out activities that use the artefact concerned. Utilisation schemes may be personal inventions, but more frequently they arise socially, through observing and imitating others, or through direct instruction. Radardel calls the combination of an artefact and its utilisation schemes an "instrument" and argues that it is instruments, rather than artefacts, that mediate human object- (goal-) oriented activity. Rabardel and Bourmand (2003) distinguish between four kinds of instrument-mediated activity:

- *Pragmatic* mediations: are concerned with action on the object (e.g. changing it in some way)
- *Epistemic* mediations: are concerned with the subject gaining a better understanding of the object
- *Reflexive* mediations: are concerned with the subject herself (e.g. with strategies for self-management, like the deliberate use of aids to memory)
- *Interpersonal* mediations: are concerned with mediated relations with other people, such as other members of a design team or collaborative learning group.

This leads us to suggest that there are four kinds of tools that may be needed in a well-furnished environment for tackling wicked problems. We can start with the *epistemic* – tools which are suited to the conduct of inquiry, to finding out more about the problem and its causes, and to helping a team create an agreed understanding. Examples would include the classic tools of research – statistical analysis programs, modelling programs, etc. If action is to be taken to resolve the problem, then *pragmatic* tools come to the fore. Individual students may also need to use tools for self-regulation – for example, in the form of a to-do list and a diary or notebook. These are candidate *reflexive* tools. And finally there may be value in using *interpersonal* tools for group management and communication: a discussion board and project management software, for instance.[5] Many powerful tools used in collaborative professional work combine interpersonal mediation with the three other kinds of mediation, such as in shared concept mapping, collaborative writing, situation awareness and other tools for heterogenous and distributed problem solving and action (Nicolini, Mengis, & Swan, 2012; Star, 1989).

Of course, the tools on their own are no use; students need to develop appropriate utilisation schemes for each of them, thereby transforming them into useful instruments. As we saw with the example of the hammer, tools can be used in a variety of ways, so it may be best to think of this four-fold taxonomy in functional terms: that is, as instruments. The key point is that working collaboratively on a wicked problem is an instrument mediated activity. The capacities needed to work effectively on wicked problems include being able to make productive use of appropriate instruments – pragmatic, epistemic, reflexive and, simultaneously, interpersonal. The utilisation schemes that turn tools into useful instruments need time and experience to develop, though direct instruction, careful scaffolding and working with peers may all be helpful in kickstarting what is inevitably a personal learning journey. From a higher education design perspective, it would not be safe to assume that undergraduate students' facility with social media and computer games would equip them with the utilisation schemes needed for successful use of pragmatic or epistemic instruments, though such experiences may give them a head start with reflexive and interpersonal instruments. In consequence, educational design aimed at helping students learn to tackle wicked problems needs to consider how students will come to recognise the range of potentially useful instruments and what is involved in jointly configuring those instruments to create a productive working environment.

CONCLUSION

In this chapter we have sketched some educational design responses to the challenge of unpredictable futures in workplaces, and in community life more broadly. We have used the idea of helping students learn to address wicked problems as a way of advancing an argument about education for problem solving. Recognising that there are different kinds of problems, with different kinds of solution methods, is an important step in thinking about appropriate curriculum arrangements, including more rational sequencing of learning activities. To be clear: we are recommending that educational design should cover *all* these classes of problems and problem-solving approaches, not just the most difficult. We are also saying that students are best helped to master problem-solving strategies through combinations of direct instruction (e.g. about specific problem-solving strategies), practical experience and feedback. While some kinds of problem-solving strategies can be learned whether the student cares deeply about the problem or not, other strategies depend heavily on students having "skin in the game" – on whether they can see themselves as inside rather than outside the system of concern. This is particularly clear with what we have called *systemic wicked problems* – the evolution of "moral know-how" can be impeded by poor choice of problems.

Our final point is to re-emphasise the importance of explicit, teachable strategies and shareable instruments in helping students learn to work on wicked problems. Successful learning does, of course, include processes through which students make personal sense of what they have experienced. They need to *appropriate* ideas, methods and instruments. This is often emotionally charged and deeply personal. It may involve processes, outcomes, feelings and intuitions that are very hard to pin down and articulate. But none of that takes away the educators' responsibilities for speaking clearly about what *can* be made explicit and for sharing useful sets of ideas, methods and instruments. Reflecting on this should stimulate fresh thinking about the interface between academic and workplace learning: about the most congenial places for different kinds of learning to occur and about the nature and distributions of tacit and explicit knowledge. Some things that need to be learned benefit from time spent embedded in the workplace. Other things are best learned through direct instruction and guided practice, especially when tacit knowledge embedded in practice has been rendered explicit through academic analysis.

Learning for work is not the only goal, of course. Indeed, we are drawn to the view that helping students learn through collaborative engagement with systemic wicked problems that affect communities and society more broadly may be a more important line of activity, and more sustaining for all concerned.

ACKNOWLEDGEMENTS

We are glad to acknowledge the financial support of the Australian Research Council through Grant DP0988307 (Professional learning for knowledgeable action and innovation: The development of epistemic fluency in higher education).

NOTES

[1] The blurriness of some of these contrasts has been scrutinised recently by Farrell and Hooker (2013), who conclude that science is more plagued by wicked problems than Rittel and Webber acknowledged and they argue that science is better conceived as design-like than design as science-like. See also Levin et al. (2012) on "super-wicked problems" such as climate change which are "wicked problems" with four additional sources of difficulty: "time is running out; those who cause the problem also seek to provide a solution; the central authority needed to address it is weak or non-existent; and, partly as a result, policy responses discount the future irrationally" (p. 123).

[2] Within the field of design studies and design research there is a growing interest in participatory, networked and community-based design of shared services and social innovation. The ideas, tools and practices emerging in this area are likely to be good sources of inspiration for those of us thinking broadly about desirable graduate attributes and curriculum design. See, for example, Manzini (2015) and Baek et al. (2018).

[3] This is an obdurate problem in the teaching of "design thinking" in schools. Taken seriously, design thinking must keep open the possibilities for substantially reframing problems – exploring their deep roots and pursuing radical solutions – otherwise it is just a bogus form of make-believe and most students will see it as such.

[4] New developments in work integrated learning (WIL) ought to offer students opportunities for participating, albeit peripherally, in work on wicked problems. Employers say they value the transformative potential of their graduate hires. Nevertheless, the emphases in WIL tend to be on lower-key workplace readiness and safety issues.

[5] Baek et al. (2018) provide a nice illustration of the value of social network analysis tools in the collaborative design of community services. While they do not refer to this as instrument-mediated activity, it is a good example of the use of an interpersonal instrument in work on wicked problems.

REFERENCES

Baek, J. S., Kim, S., Pahk, Y., & Manzini, E. (2018). A sociotechnical framework for the design of collaborative services. *Design Studies, 55*, 54-78.

Blackmore, C., & Ison, R. (2012). Designing and developing learning systems for managing systemic change in a climate change world. In A. E. J. Wals & P. B. Corcoran (Eds.), *Learning for sustainability in times of accelerating change* (pp. 347-364). Wageningen, The Netherlands: Wageningen Academic.

Brew, A. (2013). Understanding the scope of undergraduate research: A framework for curricular and pedagogical decision-making. *Higher Education, 66*(5), 603-618.

Checkland, P. B., & Poulter, J. (2006). *Learning for action: A short definitive account of soft systems methodology and its use for practitioners, teachers and students*. Chichester, England: Wiley.

Collins, A. (2017). *What's worth teaching? Rethinking curriculum in the age of technology*. New York, NY: Teachers College Press.

European Commission. (2008). *Improving competences for the 21st century: An agenda for European co-operation on schools*. Brussels, Belgium: Commission of the European Communities. Retrieved from http://eur-lex.europa.eu/legal-content/EN/TXT/PDF/?uri=CELEX:52008DC0425&from=EN

Farrell, R., & Hooker, C. (2013). Design, science and wicked problems. *Design Studies, 34*(6), 681-705.

Goodyear, P., & Markauskaite, L. (2018). Epistemic resourcefulness and the development of evaluative judgement. In D. Boud, R. Ajjawi, P. Dawson, & J. Tai (Eds.), *Developing evaluative judgement: Assessment for knowing and producing quality work* (pp. 28-48). Abingdon, England: Routledge.

Ison, R. (2008). Systems thinking and practice for action research. In P. Reason & H. Bradbury (Eds.), *The Sage handbook of action research: Participative inquiry and practice* (pp. 139-158). Los Angeles, CA: Sage.

Jonassen, D. H. (2011). *Learning to solve problems: A handbook for designing problem-solving learning environments*. New York, NY: Routledge.

Kilpi, E. (2016). *Perspectives on new work: Exploring emerging conceptualizations* (SITRA Studies 114). Helsinki, Finland: SITRA. Retrieved from http://www.sitra.fi/en/julkaisu/2016/perspectives-new-work-1

Krippendorff, K. (2006). *The semantic turn: a new foundation for design*. Boca Raton FL: CRC Press.

Levin, K., Cashore, B., Bernstein, S., & Auld, G. (2012). Overcoming the tragedy of super wicked problems: Constraining our future selves to ameliorate global climate change. *Policy Sciences, 45*(2), 123-152.

Manzini, E. (2015). *Design, when everybody designs: An introduction to design for social innovation*. Cambridge MA: MIT Press.

Markauskaite, L. (2016). Epistemic fluency perspectives in teaching and learning practice. *Epistemic Fluency*. Retrieved from https://epistemicfluency.com/2016/02/01/epistemic-fluency-perspectives-in-teaching-and-learning-practice

Markauskaite, L., & Goodyear, P. (2017). *Epistemic fluency and professional education: Innovation, knowledgeable action and actionable knowledge*. Dordrecht, The Netherlands: Springer.

Micarelli, A., Stamper, J., & Panourgia, K. (Eds.). (2016). *Intelligent tutoring systems*. Switzerland: Springer Nature.

Nicolini, D., Mengis, J., & Swan, J. (2012). Understanding the role of objects in cross-disciplinary collaboration. *Organisational Science, 23*(3), 612-629.

Organisation for Economic Co-operation and Development (OECD). (2018). *The future of education and skills: Education 2030*. Paris, France: Directorate for Education and Skills, OECD. Retrieved from http://www.oecd.org/education/2030/oecd-education-2030-position-paper.pdf

Rabardel, P., & Bourmaud, G. (2003). From computer to instrument system: A developmental perspective. *Interacting with Computers, 15*(5), 665-691.

Rittel, H., & Webber, M. (1973). Dilemmas in a general theory of planning. *Policy Sciences, 4*(2), 155-169.

Scardamalia, M., & Bereiter, C. (2014). Knowledge building and knowledge creation: Theory, pedagogy, and technology. In K. Sawyer (Ed.), *Cambridge handbook of the learning sciences* (2nd ed., pp. 397-417). Cambridge, England: Cambridge University Press.

Shaffer, D. W. (2006). Epistemic frames for epistemic games. *Computers & Education, 46*(3), 223-234.

Star, S. L. (1989). The structure of ill-structured solutions: Boundary objects and heterogeneous distributed problem solving. In L. Gasser & M. N. Huhns (Eds.), *Distributed artificial intelligence* (Vol. 2, pp. 37-54). London, England: Pitman/Morgan Kaufmann.

Peter Goodyear DPhil (ORCID: https://orcid.org/0000-0001-9903-737X)
Centre for Research on Learning and Innovation, University of Sydney, Australia

Lina Markauskaite PhD (ORCID: https://orcid.org/0000-0002-3470-890X)
Centre for Research on Learning and Innovation, University of Sydney, Australia

PAUL WHYBROW AND ASHELEY JONES

5. THE CHANGING FACE OF WORK

Considering Business Models and the Employment Market

This chapter explores the disruption to workforce models, which professionals are experiencing in the shift to an Advanced Digital Economy, better known as Industrial Revolution 4.0. Four new workforce cohorts are proposed: Emerging Learners; Millennials; Mid-careerists; and Deep Experience careerists. Through an exploration of these cohorts, an examination of the changes that are underpinning a move from a corporate silo of skills to individual career flexibility is undertaken. This chapter argues this emerging work model provides careerists with a wide approach to career foundations rather than a siloed one. This shift sees people switch from having one deep and highly experienced career to moving between different careers using transferable cross-industry skills that involve multiple specialisms. Whilst this horizontal approach can be considered a positive advancement, we need to acknowledge that the associated freedom can blur the boundaries between social and work spheres, potentially disempowering the workforce as instant accessibility becomes the norm.

SKETCHING THE LANDSCAPE

The notion of a fixed work location for most of the labour force has evolved over the 20th century from the scientific management or Taylorism approach of military-style desks through to the more open style of the 1960s and finally to the cube farms of the 1990s. The first industrial revolution, which began in the late 18th century, introduced the birth of factory employment. Fixed places of employment continued during the second industrial revolution, when the moving assembly line guided in the age of mass production. In the late 1960s, the third industrial revolution saw the beginnings of computer and automation implementation but within fixed work abodes. Indeed, a further commonality across these eras was the likelihood of a fixed form of career, often with the same employer for the duration of one's working life.

Over the second decade of the 21st century, the working environment has rapidly become known as Industrial Revolution 4.0, embedding the recognition of a fourth wave of change due to the rapid digital transformative developments that have emerged in the last 20 years. Powered by new technologies, the increased role of automation, artificial intelligence (AI), the application of big data and cyber systems, these technological advances have not only disrupted the ways in which work is conducted but disturbed the ways in which the workforce interacts with employers, colleagues, managers, leaders and customers as well as myriad other internal and external stakeholders. These two foundational shifts accelerate the need for us all to be much more flexible in the way we engage in career management, work participation and skill acquisition. The combination of an extended life expectancy and specialist skill-hopping, it could be argued, offers a very different era of personal career management.

For decades now, in Australia as in the rest of the industrial world, people have been living longer and enjoying an improved, healthy retirement. The economic reality, however, is that retirement will need to be delayed in order to fund our longer life expectancy. State pensions or final-salary related company pensions are under immense pressure, and there is a global shift to self-funded retirements. On the positive side, there is plenty of evidence that people enjoy the social value associated with work activities, which outweighs the desire to not do so. If we live longer and feel healthier then we are likely to want the stimulus of work for longer.

The second fundamental change is the need for specialist skill-hopping to provide us with the longer working span we either desire or need for economic reasons. In the shift to an Advanced Digital Economy, technology will increasingly aid our capability to operate in a deep specialist work environment. What in the past took specialist medical practitioners, commercial contract lawyers or academic professors decades to achieve, may be achieved in a fraction of the time through the combination of specialist focus and AI access to immediate global research, practices, experience and capabilities.

Longer work lives and the capability to move between specialist roles, means future workers will increasingly have a number of career roles. Potentially, with a working life of, say, 50 years, a person could easily accommodate five 10-year periods of specialist careers. These changes will disrupt fixed work sites and static career pathways. There is a recognition that contemporary careerists require different types of skills, outside of traditional discrete-discipline proficiencies to compete in the global human capital workforce. Prima facie, this fourth industrial revolution appears to allow for flexible work models, which offer a wider approach to careerists than the previous formal, vertical model. This chapter seeks to interrogate the current gig economy to better understand whether the highly hailed knowledge economy is one of freedom or constraint for four professional cohort groups operating within the 21st century employment market, that we have categorised as: *Emerging Learners*; *Millennials*; *Mid-careerists*; and those with *Deep Experience*. In turn, each of these cohorts will be examined to see where there is a cause for celebration, or concern.

EXAMINING THE PROFESSIONAL COHORTS

As past industrial revolutions evolved and expanded to the knowledge-based system of today, some groups have benefited from change and others struggled to compete. To date, the more educated, higher skilled and wealthier an individual, the more likely they have benefited. The rapid pace of change in the Advanced Digital Economy will likely mean that different generational cohorts will face different challenges. Looking at workplace flexibility through the age group cohorts of *Baby Boomers*, *Gen X*, *Gen Z* and *Gen Alpha* allows us to examine these potential changes.

Baby Boomers (Deep Experience)

We describe the members of this group as follows: they were born between 1946 and 1964 and they have had long careers in post Second World War workplaces, experiencing both massive economic growth and bouts of recession and high inflation. For most, their working world has been one in which steady innovation has improved personal productivity. In offices, for example, typing and secretarial pools have been replaced by computers and mobile devices, enhancing both productivity and career opportunities. Office memos have given way to email, text and group chat applications. In retail, the manual till has given way to automated sales devices.

Many baby boomers have specialised deeply in one silo capability. Whether an accountant, hairdresser, plumber or salesperson, baby boomers may have practised the same profession for most of their career, potentially within the same company. They received free higher education. Success was perceived as employment in a good company, and the ability to buy a home and raise a family comfortably and affordably. For those in administration, less skilled roles or ailing industries, change saw job redundancies, but luck and economic opportunities meant possibilities for new roles. Overall, educated baby boomers experienced career opportunities and relish pending retirement and increased family time.

Gen X (Deep Experience)

We describe the members of this group as follows: they were born between 1963 and 1980 and many are now at the height of their career. They entered the workforce following major developed world economic reforms: via Hawke and Keating in Australia, Thatcher in the United Kingdom, and Reagan in the United States. Through privatisations and growth in the stock market, accumulation of wealth was not only acceptable, but the norm. Gen X have spent a great deal of their careers reliant on specialisation skill sets. For some, investment in property provided wealth, which grew faster than wages or salaries. Gen X were exposed to far more to digital transformation than previous generations. They are still a decade or more away from retirement but may be realising their superannuation balances are less healthy than required. Real wages have risen slowly, and Gen X have not been prudent savers.

Easily accessible credit has allowed this group to spend beyond their earnings, on such things as overseas holidays, new cars, flat screen TV's, smartphones and the delights of modern Australia. In the workplace, experience is less valuable than it once was. Hierarchy and status are being replaced by flat structures, open plan offices, agile projects and consumer-centric technology. As this group begins to enjoy the status of seniority, many roles are under threat from the AI revolution and this group of specialists is not equipped to handle massive career jumps.

Simultaneously, higher pension ages have been brought in as the Australian Government and companies look for ways to fund the retiring Baby Boomers.

Gen Y (Mid-careerists)

We describe the members of this group as follows: they were born between 1981 and 1994. They are in mid-career. The combination of extended life expectancy, rising age of access to state pensions and the demise of final year salaried pensions mean that this group will need to work longer than earlier groups to achieve their desired retirement income. This group has experienced increased costs of housing and childcare, and many have also incurred debts for higher education.

Financial pressures are much higher than on baby boomers, with two incomes often being required to meet obligations. To adapt to the multi-career environment, mid-careerists may have to undertake second or third careers to support their retirement. With the introduction of digitally transformative jobs, they will need to stay ahead of the Millennials and Gen X. In their childhood, they may have watched DVD's and listened to CD's, but in adulthood they are the frontline for the digital working environment, pioneering the adoption of new flexible working. They have been able to do this with confidence as economic prosperity has been good, with no deep recessions, and until now their chosen careers have felt secure.

Gen Z (Millennials)

We see this group as a digitally confident generation who only know life through Google, Facebook, Netflix, smartphones, texting and social interaction that is dominated by devices. They are driven by two very strong passions: *convenience* and *experiences*. They have gone through school and higher education in a boom time, adjusting to a consumer-centric personalised world. They search online using Google rather than read books, use navigation applications rather than asking for directions, and love the idea of obtaining transport, entertainment and food anywhere and at any time using a mobile device and virtual money. This group also expect this level of experience and convenience in their working careers. As Zimmerman (2018) explains, Millennials are very confident because they are used to researching prospective employers on review sites and social media and so have more confidence on what to expect when they change employers.

In some ways, the world of start-ups and the gig economy seems very tempting as a flexible and fulfilling way to continue exploring what Millennials want to do. Unlike the generations before they are not so confined to a pattern of career and rapid home buying. This group appears to love flexibility, "chilling out" and taking stock. And yet there are signs that this group have their own flexibility challenges, such as building the resilience to cope with unplanned change or dealing with the unfairness of the economy, shifting consumer tastes or the capability to navigate the critical thinking, traditional communication and leadership skills needed to engage successfully with the consumers of the earlier working groups.

Gen Alpha (Emerging Learners)

Born since 2012, we describe these people as the future workers of Australia, who have only just started primary school. The rapid digital transformation we are seeing now in the workplace will permeate through to the education practices and experiences they will have. Some futurists suggest they will be the first post industrial revolution generation who will never learn to drive, they may always have a digital assistant to sort out their life, and they might never need to learn facts at all, as everything needed will be available instantly wherever problems arise. Work for this generation is likely to be vastly different to what we know today; whether it will be the thing of real science fiction is hard to tell. One key likelihood is that Gen Alpha will be working and studying for longer, perhaps not reaching the workforce until their mid-20s and then not stopping work until their 70s and 80s. That 50–60 year working life will likely require a great deal of skills and professional flexibility.

CONSTRAINT VERSUS OPPORTUNITY

Rapid technological change that drives massive change in society is not new. For example, the first industrial revolution saw the shift from workers living and working on the land to working in factories and offices. The technology of the steam engine and the production line was the foundation for the modern society that provides Australians with good education, political freedom, a world-class health system and high personal living standards. In contrast with today's workplaces, many workers during the first industrial revolution faced long working hours, child labour, unsafe work practices and a culture of complete self-responsibility for safety. Periods of workers fighting injustice and winning social reform have driven the path to our current set of rights and expectations of fairness, equality, safe work environments and sustainability. With the emergence of the fourth industrial revolution and the growth in life expectancy, the key question to be asked is whether the Advanced Digital Economy will lead to *greater freedom or a return to serfdom*.

FREEDOM VERSUS SERFDOM

Social and economic transitions are underway, which can impact upon professional cohorts, depending on the individual or collective ability to benefit or lose from these changes. We are already seeing tensions build around the fear of the changes that could occur. For instance, the ubiquity of mobile devices, leading to an always-connected environment, has massive personal benefits for connecting to others and/or obtaining information from anywhere in the world, but many people are struggling to escape 24-hour connectivity with work and the need to answer every email or text. The so-called "gig economy", where people are free to move from project to project wherever their skills and desires take them, seems appealing, but the increase in contractors and sessional staff means a significant portion of the working population does not have full employment rights, lacks capacity to save for the future and fears periods of unemployment.

With increasing capabilities of AI and robotics, many roles we know today might disappear, especially in the skilled professions, which until now have been unscathed in the shift to digital technology. This could generate an underclass of people who have the flexibility to shift careers, but are unable to adapt because of lack of training, transferable skills or economic pressures. Gen Y, who often need years of income to manage growing families and save for retirement, could be especially vulnerable to the cycle of falling jobs, deflating wages and high competition for remaining or new jobs. In this case, flexibility becomes a burden required for survival rather than an exciting opportunity. Clearly there are massive differences to the harsh and often brutal capitalist conditions at the end of the 19th century. We have access to knowledge like never before, we are healthier and live longer, can live in smart homes, travel the world and have opportunities for leisure activities and entertainment that a Victorian worker could not imagine in their wildest dreams. We could, however, be on a path that is heading to the illusion of flexibility, which reduces freedom rather than enhances it.

OPPORTUNITIES FOR FLEXIBILITY AND FREEDOM

Opportunities created by technological changes and longer living will impact all generations to a degree. Five years ago, who would have believed that: an application for car owners would disrupt the taxi industry; Netflix in the US would become one of the biggest producers of content globally; or professional network platforms such as LinkedIn would be readily accessible on our mobile phones? Although predicting the future is impossible and unwise, there are some likely trends, and some seeds of innovation that look like they will continue to flourish and grow. These four flexibility drivers are available to all generations: *location*, *convenience*, *experience* and *knowledge*. The adoption of each of these drivers within the generational cohorts will be dependent on the ease and access of technological uptake.

Location

It is likely that, at some point in the future, the ability to run your work entirely from your mobile and laptop will become possible. Assuming that high speed connectivity will be achievable everywhere, then in most cases, any employee or small business will have access to all corporate systems, customers' information, account systems and transactional services. There will be no longer be a need to go into an office or fixed location. Whether that means all office environments will go, is less likely. People use fixed work locations for several reasons.

For retail, warehousing and physical services and entertainment, a fixed location makes economic sense. For technology, creative and professional services it doesn't have to, except for human nature. Humans are social animals and often we can be more creative as a group, together, than as individuals, at a distance. This might be because we like the company of others, benefit from encouraging one another, or require distance between locations for work and play. The growth of start-up collective work spaces suggests that centralised and shared office accommodation might not disappear overnight.

Convenience

Convenience will drive expectations around how individuals work together and interact with customers or clients. Electricity, the motor vehicle, the washing machine, telephone, newspaper, radio, TV, the Internet, industrial and home automation: all have made our lives easier. Each generation strives to make life easier, taking inherited convenience as the norm. In the future, it is likely that almost anything could be delivered instantly as required. Computing technology is likely to increasingly replace hard labour and routine administration. This, in theory, will allow us more flexibility with our time. Intensive craft and labours of love are likely to become hobbies that are chosen, not endured. Convenience living could lead to ever increasing use of online access for whatever you desire in life. Whether it is shopping, finding a partner or ordering take-out food, all of these activities are already very convenient, and may simply become the norm.

Experience

Creating and delivering a personalised and convenient experience is a clear focus for producers. Individual experiences are likely to be increasingly provided to customers using AI and other immersive technologies. Also, it is likely that workers will want individualised work experience to match where they are in their career and life stages. For example, a mid-20s worker may want to advance professionally, while exploring, discovering and enjoying their work experiences around a work hub. They won't mind working and playing hard, with social life on tap, and want work life to match this. For many, time efficiency won't be key. This might contrast with someone who is raising children, for whom efficiency at work will be key. They will likely want the information to do their job and far less internal interaction with colleagues. They may want more home-based and mobile activity. They probably will want work to support their family and caring focus. The future workplace environment might, therefore, have to deliver varied experiences for workers and customers, compared with current workplaces, which often expect to employ many people doing the same role.

Knowledge

Access to knowledge that is constantly updating to be incredibly contextual will power most workers' flexibility, including aforementioned "specialist hopping" and multiple careers in a working lifetime. That lifetime might spread over 45 or more years and include: higher education (5 years); travelling and discovery (2 blocks of 2 years); mid-career retraining (2 x 1 year); staying home working part time to focus on bringing up young children (7 years); and three periods of deep specialism with different professions and companies (30 years).

The fuel to drive this flexibility will be the ability to tap all global knowledge in an instant and to layer specialist human expertise with exponentially growing AI and robotics power. It is expected that program and content updates can occur without having to think about it; if you are subscribed you automatically get it all. It will be accessible from anywhere via any device: just as your music can switch from your phone to your car seamlessly at present, so will future access to individual and business systems and knowledge.

It is this capability that will allow people to become strong experts very quickly and then operationally always have access to the latest trends, research and consumer sentiment. Once this is married to deep and live business knowledge for your area of specialisation, along with live data and analytics, personalised behaviour, live pricing and delivery or service capability, then individuals will have the freedom to access leading global knowledge and expertise. The specialist won't be replaced by the technology or machines; however, it is increasingly likely that they will become smarter and more able to meet the increasingly individual desires for convenient and positive experiences in all types of work activity and goods and service delivery.

Deep Experience

These people are at the point in their career where they are likely to be eyeing their retirement prospects and comprise two groups who have different career motivations. Those in the first group who have stayed in one career most of their working life without having been heavily involved in technology change and who want to just work another few years before retiring, are likely to be vulnerable to change, including replacement by AI or commercial restructuring. They are unlikely to have the skillset for the evolving workplace where experience may not be valued as in the past. They don't have the tools or time to reskill and benefit from what flexibility might offer.

Members of the second group have been close to technology and business change and/or have varied career roles. They have built resilience and thirst for change, may be unburdened by cost as their adult children head into work or tertiary education, and will shortly have access to a good superannuation fund. Embracing the available flexibility, they may step aside from senior higher pressurised roles, or start their small business, using their experience and skills whilst pursuing travel, family time and hobbies. This group may be comfortable to work in multiple locations in offices, shared spaces or in cafes. They can use mobility to advantage and tap into their strong network to steer short-term career options.

Mid-careerists. This group are probably the most vulnerable to the rapid change that AI and flexible working could bring. Many may have several financial and social constraints. Mortgages and costs of bringing up children along with split family units add strain and so the desire for stability could be very high. Many previously unchanged professional roles may be swept away by AI, machine learning and the new needs of the digital consumer. If active steps are not taken to retrain or develop transferable creative leadership and soft skills, then this group might be unable to adapt and reskill quickly enough. Affected individuals may require government-backed education or strong company outplacement support to re-skill.

Millennials. Convenience, flexibility and consumer experience are second nature to this group. Their digital experiences and opportunities have given them the spirit and the capability to risk jumping from career to career. They are used to mobility and mixing work with personal activity, and so can take advantage of all that the new future offers. Whether they are in the gig economy, fixed contracts or full-time work, they will keep adapting to the next wave of technological change and will be happy if they get adequate periods of family life and travel along with career advancement.

Emerging learners. Gen Alpha are the hardest to predict because they are years away from entering the workforce. They are digital through and through and will go through their education demanding digital over physical in many things they do. By the time they reach the workforce a multi-career lifestyle is likely to be the norm. They are likely to take their time: a single degree may shift to a double degree, and these may be interspersed with short periods of working or travelling. They can look forward to 40 or maybe 50 years of earning capacity and will expect all of it to be a good experience.

They will have access to much knowledge and an array of new and as yet unknown careers. Their life will be very different to that of their grandparents. They will expect that they can do anything they want, with complete fairness and equality and no global barriers. The mechanics of work are likely to happen in the background with little or no human interface. Creative leadership, smart application, openness for change and team problem solving may be the key attributes that will be taken from role to role. Technology improvements and true flexibility to do whatever I want, where I want and often when I want will mean that work will be fully integrated into life in a very comfortable way.

CONCLUSION

This chapter has explored the disruption to workforce models which professionals are experiencing in the shift to an Advanced Digital Economy. It reviewed the modes of disruption affecting the workplace through the lens of specific demographic cohorts. In reviewing the various themes identified as impacting the work of the future it can be argued that these changes could be viewed in terms of negative impact, unless the economic social elements are better addressed. If the socioeconomic concerns identified are interrogated, then the potential future of the workplace environment could positively impact on the professional careers of the future.

REFERENCES

Zimmerman, K. (2018, February 11). This is why Millennial job seekers are so confident. *Forbes.* Retrieved from https://www.forbes.com/sites/kaytiezimmerman/2018/02/11/this-is-why-millennial-job-seekers-are-so-confident/#722934891eb4

FURTHER READING

Monahan, K., Schwartz, J., & Schleeter, T. (2018, May 1). Decoding millennials in the gig economy: Six trends to watch in alternative work. *Deloitte Insights.* Retrieved from https://www2.deloitte.com/insights/us/en/focus/technology-and-the-future-of-work/millennials-in-the-gig-economy.html

Press, J., & Goh, T. (2018). *Leadership, disrupted: How to prepare yourself to lead in a disruptive world* (White Paper). Greensboro, NC: Center for Creative Leadership. Retrieved from https://www.ccl.org/wp-content/uploads/2018/01/Leadership-Disrupted-White-Paper.pdf

Roser, M. (2018). *Life expectancy.* OurWorldInData.org. Retrieved from https://ourworldindata.org/life-expectancy

Paul Whybrow
Varda Creative Leadership and Bodyboard Immersive Experiences

Asheley Jones (ORCID: https://orcid.org/0000-0002-2441-6321)
Zena Consulting, Australia
Future Work Initiatives, Deakin University, Australia

PART 2

PRACTICE AND THE COMMON GOOD

DEBBIE HORSFALL AND JOY HIGGS

6. RE-CLAIMING SOCIAL PURPOSE AND ADDING VALUES TO THE WORLD AROUND US

FIRST, A RANT: THE GOOD, THE BAD AND THE ANGRY

The Western obsession with productivity has brought the world to a crisis that we can escape only with a radical break from the headlong rush for "more, always more" in the financial realm as well as in science and technology. It is high time that concerns for ethics, justice and sustainability prevail. For we are threatened by the most serious dangers, which have the power to bring the human experiment to an end by making the planet uninhabitable. (Hessel, 2011, p. 19)

We are rather angry that we have to write about re-claiming social purpose and adding value to the world. It is, to say the least, distressing. It disrupts and challenges our beliefs in the emancipatory agenda of the educational project and it destabilises any ideas of having made some sort of contribution to "the better society". It speaks to the fact that we may have failed, or at least not worked hard enough. Our whole sense of purpose has been to make a difference, to pursue social good, to add value not only to the world around us but to the people around us. A rather grandiose notion you may say. Possibly. But it has been the ethical and moral compass of our work. We share the moral outrage of Hessel (2011). Perhaps it is time for such outrage?

We are not claiming that we are the only ones, that would be both arrogant and incorrect. There exists a community of scholars and professionals challenging the way things currently are, critiquing the current order in their workplaces, in their professional bodies and organisations, and in their private lives. While such critique has been a traditional part of both the professional and academic worlds it appears to be becoming less inherent and less pervasive. Indeed, critique has become increasingly an external and imposed part of our lives and work rather than a normal and expected part of what it means to be a professional. How many times is our work disrupted by accreditation agencies, government authorities or organisational leaders who have developed yet another system of regulation, review and monitoring of work goals, purpose and outcomes? We see that making people be *other*-regulated instead of primarily self-regulated, actually reduces opportunities for and expectations of self-regulation and reduces the pursuit of social good, which is a primary criterion of professionalism. Why bother trying to check our collective systems and practices if others are constantly making us do it to their agendas and often concurrent and conflicting standards?

We observe that those, like us, still trying to pursue social justice are in the minority. We have become the Other not the norm in professional practice. We are the less dominant ones, writing counter narratives, thinking counter thoughts, researching marginalised stories and voices. Perhaps we are the resistance? And, of course, as two white, educated, employed, middle-class academics we are also not the resistance: some may say that people like us are part of the problem (Moreton-Robinson, 2000) as we write from within the halls of power. Although we inhabit these spaces and places of privilege our practices are ones which work to unsettle, question, probe, subvert and support differences. We aim to use our privilege to make our profession, our workplaces and our worlds more democratic, inclusive and kind as we operationalise the notion of strategic privilege: using our positions for collective benefit as well as making sure we are taking care of ourselves. It is certainly an ethic of care where "[t]he ethic of care teaches society that it must act when it can, and that the moral conundrum is not *whether* to act, but *how* to act" (Power, 2002, p. 115).

And isn't this also the point of professional practice? To take actions, big and small, which make the world a just, better, beautiful, more caring place for all of us? Police to keep us safe and tell us when we are doing wrong; nurses to care for us when we cannot care for ourselves and each other; doctors to make us better when we are sick or injured; architects to design functional, safe and beautiful places for us to live and work and recreate; teachers to help us grow into competent, creative and knowledgeable people who can co-exist with others and contribute to the world; scientists to help us make sure we still have a planet to live on? In the absence of these ideals and expected practices of professionalism we find that – rather than working for the greater good – regulators manage systems, organisations lead national and global social and economic plans, companies generate income, and professional employees are no longer taught or expected to repay privilege and benefit with service to society. We would like to be writing this from a different place as we imagine the future. We would like to be in a place where values have not been thrown out the window. However, they have been and this makes us annoyed. As Hessel says:

> It is up to us, to all of us together, to ensure that our society remains one to be proud of: not this society of undocumented workers and deportations, of being suspicious of immigrants; not this society where our retirement and the other gains of social security are being called into question; not this society where the media are in the hands of the rich. (2011, p. 15)

Not this society where a woman a week is killed by an intimate partner in Australia, where we have to hold Royal Commissions into black deaths in custody, the banking sector, the abuse of children by institutions meant to be caring for them, where numerous reports tell us about the appalling numbers of students who experience sexual abuse and harassment in our universities, where we are told that we can no longer as a country afford to be decent, to pay liveable wages to everyone, to stop the alarming extinction of species. Where, to our shame, we lock up people escaping war, famine, violence and persecution; where we are closing women's shelters and not putting enough resources into child protection; where we are told we have to fund our own retirements even though for many this seems like some cruel joke and the brick worker's labourer may have to continue their backbreaking work until they are 70! And, where many people accept that this is the way things have to be.

You might be wondering at this point who we are talking to. On one hand we ask all of our society to be listening. On the other we are talking to the privileged and the recognised leaders and elite. We are indeed talking to those who claim to or adopt the mantle of leadership and the practice and responsibility of politics, society sponsors or patrons, those in authority in industry, communications and commerce, the intellectuals and the educational leaders. While each of these groups do not fit the role precisely of membership of the professional class, they do lay claim to the title of community leader and therefore need to recognise, honour and accept the attributes and practices of professional people.

SO, HOW DID WE GET TO THIS POINT AND WHY IS UNDERSTANDING THIS NECESSARY?

In one of the first classes of a subject called *Sustainable Futures* at Western Sydney University[1] students do a futures exercise. Framed around probable, possible and preferable futures they are asked to imagine how things could be different. To get them started, the example of free university education is used, with the lecturer saying that for him this is a preferable future to one where people are paying ever increasing university fees.[2] "Let's start the free education movement," he says. "Who will join me?" About 50% do; the other 50% will not. As background, this subject is run in the western suburbs of Sydney, a relatively disadvantaged area with most students being first in family to attend higher education, and most studying and working with an increasing number working full time and studying full time. When asked why they will not join the movement their answers fall into two categories: 1) as a country we can't afford it; and 2) people only value what they have to pay for. From these answers the class moves on to discuss the purposes of higher education and related considerations of state versus private responsibilities for its funding. The goal is to help students learn to question the taken-for-granteds in society and how this links to sustainable futures and responsibilities.

Extrapolating the above exercise into this chapter we ask: how can professional practice be different in the future? The title of this chapter is a bit of a spoiler – we believe that we need to re-claim social purpose in the professions. This re-claiming encapsulates our vision of a preferable future, of how things need to, and can, be different. Of course, imagining and working towards a different future asks that we are also able to understand the present by engaging in a process of deconstruction so that we can reconstruct alternatives, or at least to leave space for alternatives to flow into. It also asks that we do not uncritically accept the visions of the future being offered to us or expected of us. In terms of public and political policy the future is depicted by some "as a globalising march towards 'frictionless capitalism' based on information and communications technology, a global market from which individuals may profit but to which they must unquestionably submit" (Sullivan,

2004, p. 15). As we write, for example, a global conversation about free trade is taking place with an almost hysterical call to kill off any form of national protectionism. Indeed, this march towards frictionless capitalism is presented to us by our political elites and many sections of the media as a good thing, something from which we all will indeed benefit, with the move (back) towards tariffs signalling the end of the world as we know it. We, along with many others, want to question the lack of social justice and equity underpinning such practices and blind allegiance to them. Many would argue that it is not capitalism, markets or economics per se that are the problems but the regulatory and/or incentives structures that prevent social values being recognised in our markets.

The professions occupy an interesting place on this march. While not often spoken of as such, they do still operate as collectives and, as evidenced by codes of conduct, ethics protocols and (some) individual practices, and they still have at their core values which speak to contributing to the development of a "good" and "better" society through embracing the spirit and practice of public service and *social* partnerships (Sullivan, 2004). It is interesting to reflect on who wants to claim this title of professional and which occupations pursue this status. In the past "society" granted the privilege of professional status to occupations that had earned their entry to the professional class by meeting key criteria (self-regulation, accountability, a recognised body of knowledge, operation under a code of conduct, completion of degree-based education). The traditional model of professionalism emphasises autonomous expertise and the service ideal. In his classic paper *The Professionalization of Everyone?* Wilensky (1964) explored issues associated with trends towards the widespread pursuit of professional status by many occupations. Factors influencing this trend include the goal of achieving "the knowledge society", the expansion of knowledge to the point of needing and desiring specialisation of occupations and knowledge, expectations of standards of performance and service delivery, and market competition with professionals demanding higher remuneration, autonomy and status.

In recent decades we see a greater *claiming* rather than *awarding* or *earning* of professional status. This brings with it contradictory outcomes: status reward without quality commitment, proliferation of "professions" with less pursuit of original purpose and obligations (including service to the community) and increased apparent power but also increased state regulation and demand for the required external accountability typical of globalisation.

In the collective understanding of the professions the individual professional has a civic identity which is at the fore. In tension with this is the rise of the increasingly individualised privately employed professional who possesses expertise and provides expert services (Sullivan, 2004) to so-called consumers of professionals' services. This professional does not have a civic identity, or at least not one that is central and valued. This unspoken collectivism of the professions poses a direct threat to frictionless capitalism and sheds some light on the neo-conservative agenda to destabilise and individualise the professional classes. There is little doubt that there is indeed a crisis in the professions in Western democracies (Fook, 1999; Rossiter, 1996). As Beck and Young (2005) argue:

In recent decades, professions and professionals have faced unprecedented challenges: to their autonomy, to the validity of any ethical view of their calling, to their relatively privileged status and economic position, and to the legitimacy of their claims to expertise based on exclusive possession of specialized knowledge. (p. 183)

In a large part this crisis in the professions is due to the changing contexts of practice but also to how the professions have responded to, ignored or failed to adequately deal with changing social and practice contexts. In addition, this crisis is due in some part to increasing client autonomy and empowerment, increasing use of technology and well-informed citizenry utilising the knowledge commons available on the world wide web (East, Stokes, & Walker, 2014; Kreber, 2016). Trust in the professions has been eroded over recent years and the professions have been accused of being arrogant, self-serving and paternalistic, operating as a closed shop, mystifying knowledges and practices, and serving their own interests not the public good.

And to some extent this is true. In a recent research project, for example, exploring lived experiences of caring for people at the end of life, one of the major findings was that the medical and palliative care professions were indeed paternalistic at times putting the profession first (Rosenberg et al., 2017); it was also found that this had occurred because there was no rigorous functioning social partnership between the professionals and the public they cared for. Academic critiques of the professions say that they have always been as much motivated by private gain as the public good, and their elitist, exclusionary and inequitable practices have been challenged especially in respect to class, race and gender (Martimianakis, Maniate, & Hodges, 2009). Indeed, Charles Dickens' novel *Bleak House* is, in part, a critique and commentary on the legal profession in the 18th century, so this is not a new problem. However, it has gathered pace over the past 30 years (Apple, 2006).

In terms of this chapter what we find most interesting about this so-called modern crisis in the professions is that the professions have, on the whole, responded to the crisis by moving further towards individualism and self-interest and embracing market principles in the operation of professional practices and organisations rather than acting to further strengthen the collective power they can use for the public good, thereby pursuing more strongly the value of social goods and social partnerships. Part of this change in attitude of the professions comes from the "market place" mentality and practices of the professions and at least part of it on the shaping of those who enter the work arena by its own market place cultures.

THE ROLE OF THE UNIVERSITY AND THE ACADEMIC PROFFESSION

Part of our concern in writing this chapter is what education is doing about the future of work and the sense of direction our university graduates and our school children are gaining. One message that we hear in the education sector is "that is what students want". And so students with the power to "buy" education "vote with their feet" in choosing what university to "invest" their enrolment dollars in are shaping the education sector. But how much do they realise that what they are getting back is a commodification of education? They choose a commodity, then it is packaged, shaped and charged back to them. Perhaps for some this is the ideal future; perhaps their education has just become the slickest marketed package. In this process we see the popularity of top market approach options over integrated programs, and recognise that they bring both benefits (e.g. accessibility, particularly for part-time workers) and problems such as packing up learning bits and pieces with limited university responsibility for the whole program or the shaping of novices' induction into communities of practice. It is little wonder that many of today's graduates focus on their individual priorities (income, career paths) over their clients' needs; that is what higher education is rewarding them for choosing. And, if some commodities (courses, modes of participation) appear to be the most popular, it is understandable why universities are quick to pounce on new trends with limited critique of the cost to institution, community and students. Neoclassical economists assume that the world will be fine so long as people are able to make rational decisions in their self-interest. Putting aside the self-interest aspect, it is clear that some students are not sufficiently knowledgeable to be able to make a choice that will give them the best individual outcomes, not to mention the best social outcome. A key adjustment required in education marketing is the provision of sufficient, accessible and relevant information for prospective students to make informed educational choices.

The academic profession plays a role in influencing other professions, and its own, as well as influencing future generations through higher education. Consider this story of Debbie's experience.

In 2018 my workplace asked me to do a research leaders course. The course comprised researchers from universities around Australia and we met for six workshops over the course of a year. Prior to the first workshop we completed an online standardised team management profile about work styles and preferences. We received the results at the first workshop. A number of people were a bit upset about their low scores on the values and creativity metrics. What that means is that in terms of leadership style, decision making and work preferences they were not values driven, neither it seems were they creative in their decision making – according to this exercise. What surprised me was not that I scored highly on values and creativity, but that I had thought most of us would. While I was surprised at how far I fell outside of the norm, it explained, albeit in a simplistic way, what I observe in the university sector where compliance, metrics, rankings, marketing and the corporate university are in ascendance. These things affect and shape people's behaviours and beliefs. They change who we are and what we do – that's the point of them. And here was the evidence right in front of me. Good people shocked at the fact that they no longer placed values front and centre of their work. This exercise helped me to further understand the people around me especially the decision makers who are, it seems, quite normal in today's university by not being driven by collective notions like social purpose. They are trying their best to survive and perhaps thrive, in a particular environment which also does not value social purpose. Despite the rhetoric this is not what they 'measure'.[3]

Consider this second set of cameos from Joy's work with a series of university-wide programs that involved a mix of external quality regulation and internal participation of different styles and motivations.

A leader – a real statesman – approached his task to draw the university into an external review by saying – yes we have to do this, but let's not see it as just an unavoidable and unpleasant task. We can take this opportunity to choose to do what we do better. Let's turn compliance into constructiveness. Another person – in a position of leadership – took many opportunities to remind people of the need for each person to complete their KPIs (key performance indicator measures) – so that their collective achievements helped him to gain his own measures of satisfactory performance.

At a later occasion one team responded to a call for reflection on how well they were achieving government-imposed standards by saying – 'here's the data we collect every year to record our standards compliance'. Three other groups used the requirement to each write a performance report against government standards, to work together to share strategies they had developed to improve their work against the standards.

Willingness to pursue improvement voluntarily rather than being driven to minimal and obligatory compliance is still alive and well – as long as we make space for it be realised and avoid drowning ourselves and others in overwhelming volumes of busy work and un-mitigating change agendas.

Universities and individuals (students, academics and professional staff) have been increasingly produced by the Western neoliberal project. This very successful social change project has promoted and rewarded individualism and fragmentation of the social world and social interests with narrow income-driven market economy settings and standards in many cases providing the main set of criteria by which to judge what is good and successful. The individual is at the centre of this project: the neoliberal subject (autonomous, free, unencumbered, rational, existing outside of history and social context) is, at best, an illusion. (See note below on an alternate economic perspective that more socially oriented organisations are adopting.)[4]

We are turning our attention to the nature and role of the university sector for two reasons: firstly, as insiders we have witnessed changes in both the work of academics and the nature and function of the university sector over an extended period of time. Secondly, members of the professional classes are educated at universities. We have been affected by major university restructures in Australia firstly initiated by the Dawkins education reforms in the late 1980s (with similar reforms in many other Western democracies) which has resulted in a rise in the number of people attending university, a decrease in public funding for the sector and the rise of increasingly neoliberal and managerialist policies and practices. In our day-to-day lives this has meant larger classes, an increase in the use of technology in teaching, fee paying or partial fee paying students, students who are increasingly both studying and working full time, increasing transitory work via the casualisation of the teaching workforce, and pressure to bring money into the workplace in the way of large competitive research funds, among other things. It could be argued that the university sector and the academic profession are also at crisis point. We can certainly argue that things have changed in terms of social purpose, the common good and university's role.[5]

Until recently universities were viewed as institutions which were necessary for the common good of society, and society provided funds (via taxes and government funding) for the institutions to function and for people to attend. Graduates provided the professional services that the general public, or society, needed in terms of health, education, public service, law and order, and governance. Research was similarly supported because of its potential social benefit. Graduates did of course individually profit from their university education, but this private benefit was generally framed within the broader notion of their public role.

Historically then, universities were seen to be valuable public institutions in the service of the common or public good and they were valued for their public role and social purpose. They provided places and spaces for students to engage in robust, thoughtful and informed discussion with the aim to produce responsible, highly skilled and competent, engaged citizens. At the same time universities and the academic profession did not escape the critique of the professions overall. The public became less tolerant of perceived academic privilege and self-promotion and there was a decline in public trust of the institutions (Hazelkorn & Gibson, 2017). Universities were criticised for being elite and exclusive protected ivory towers removed from their local community's interests. This criticism was one of the reasons given for the large structural changes that have taken place in the sector. Universities needed to become more inclusive and relevant and needed to produce graduates and research which more closely and clearly contributed to society. And we would agree with those sentiments.

However, the place we find ourselves in is a complex one: mass education of course can be good for society and it is certainly more inclusive and democratic than 100 years ago when mostly wealthy white men attended. Today, many university communities are vibrant, multi-racial places with policies and practices to include and celebrate sexual and religious diversity, for example. At the same time, it is also true that while universities have been producing more and more graduates over the past 30 years, social injustices continue with social inequality increasing in many Western countries. The environment is also in crisis and it is unclear if and how the human race will survive. And this is the paradox: we desperately need people with knowledge, skills, imagination and creativity to help us negotiate our way through the combined crises of our times and

> it is the people coming out of the world's best colleges and universities that are leading us down the current unhealthy, inequitable, and unsustainable path ... making us question whether the professionals we rely on really have our, and the world's, best interest at heart. (Kreber, 2016, p. 124)

We would argue that instead of democratising education with the aim of contributing to the common good, the effect has been instead to contribute to the economic good (with the underlying assumption being that this will translate into the common good) and the notion of frictionless capitalism we spoke of earlier. This has happened as the debate and policy about higher education has been systematically hijacked by dominant groups in society who have a different aim than the common good, or at least a different conception of what the common good means. For these groups the preferred future is one which values "traditionalism, standardisation, productivity, marketization, and economic needs" (Apple, 2006, p. 22). In practice this has meant that over the past 30 years of educational reform the major focus has been on strengthening the ties between "education and paid work and education and the market" (Apple, 2006, p. 23). This has resulted in the overall rationale for "getting an education" being to serve the needs of the economy, not the community or society as a whole. Here students are positioned as so-called autonomous individuals who invest time, money and effort into becoming educated with the overall expectation that the benefits of this effort should accrue to the individual. Getting an education is seen as a personal and private investment to get a job, preferably a good job in the professional classes. Indeed, this was stated as such in a recent working paper from the Centre for Global Higher Education:

> The importance of responding to labour market needs is not simply acquiescing to the market, but responding to the needs of students for employment. In the wake of prolonged recession and slow recovery, there is an expectation that higher education, given its importance to society and the economy, and to individuals, has a responsibility to help meet these needs. (Hazelkorn & Gibson, 2017, p. 16)

The logic is hard to argue with and that's because to a certain extent it's correct: most people need an income to survive and most income is obtained through employment. However, notice how the needs of society and individuals are framed within an economic imperative and context, not a social, cultural or environmental one. They are not irreducible or even always complementary. Societies, individuals and universities are complex, diverse and emergent – so which bits of society and which individuals are we talking about? The elites. Those with power and influence and money. What is silenced, ignored or brushed under the carpet, left on the margins, excluded? Whose and which bodies, ideas, feelings do we dismiss or disregard? Which species will we let die, or kill? What island will we let drown? Yesterday the last white rhino in the world died. Knowledge can be used for the "public bad" as well as for the public good (East et al., 2014, p. 1629); not all jobs are good for us and the planet, nor is rampant economic growth. Imagine if the recession mentioned in the above quote had led to us re-inventing living well with less, for example.

In this world of knowledge capitalism (Olssen & Peters, 2005) and knowledge economies, vice-chancellors are constructed and perform as CEOs of for-profit corporations with many academics also accepting that personal, private investment and immediate financial returns are the new source of public good (Thornton,

2015). One noticeable effect of this knowledge capitalism and marketisation of education can be seen in the rewards given to vice-chancellors with Australian university bosses being some of the highest paid in the world – who interestingly, or just predictably, are mostly white males (in 2017 in Australia women comprise 25% of these roles). In many cases the salaries of vice-chancellors mean that they take home more in one week than a casual academic employee might earn in a year with men earning significantly more than their female counterparts. Is there some relationship between the increasing salaries of the vice-chancellors and the increased casualisation of the academic workforce? Lyons and Hill (2018) argue that "[t]hese inequities – of which salaries are part – constrain university research and education in the service of the public good" (n.p.).

Of course, we are not necessarily suggesting that this is how the university elite sees things – indeed many may be dismayed by what we are saying, believing that they are indeed pursuing the common good and that universities are pursuing a social purpose. Others may feel that there is no alternative and it's all about conforming to survive. However, some of these elite did write this:

> Universities make an essential contribution to creating a more diverse, sustainable and vibrant economy with opportunities for better jobs and more fulfilled lives and, through research and innovation, the creation of new products and industries. (Universities Australia, 2013, preface)

So, we are left with some questions about the role of universities, the professions and social purpose:

- What is meant by social purpose? Is this what people in society are asking for? Is it aimed at today or the long term?
- Is education a civil right and a public service rather than a commodity-based world where degrees are exchanged in the market place?
- Is the role of teaching and research to contribute to creating a world worth living in for the many not just the few?
- Are universities educating global democratic ethical citizens or citizens for a neoliberal white Western view of capitalist democracy?

Universities are actors in this space: they have agency and power. At the moment it feels as if they are doing the work of the neo-conservatives and ignoring alternatives to their own increasing corporatisation and service to the market. However, it does not have to be this way. Unlike Australia, the US and the UK, there are over 24 countries across the world including Finland, Norway, Germany, Greece, Malaysia, Morocco and the Czech Republic that continue to provide free university education, often including education for overseas students, with some universities providing a basic grant or living allowance. South Africa is currently in the process of providing free university education to their citizens with the overall aim to tackle the country's poverty (Head, 2017). Such examples challenge the TINA principle (there is no alternative – a term credited to Vandana Shiva) and show that there is not one trajectory available, nor one future that is possible.

RESISTANCE AND "RE-CLAIMATION" THROUGH CIVIC PROFESSIONAL PRACTICE?

> Futures studies seek to help individuals and organizations better understand the processes of change so that wiser preferred futures can be created. (Inayatullah, 2008, p. 5)

So far we have discussed, and offered an explanation of, what is currently happening in the professions and suggested the future which could automatically flow from this. (See Chapter 1 for further discussion of professional futures and roles.) However, it does not have to be this way. In looking for alternative, and wiser, futures a good starting point is to find and make visible stories of resistance and alternative practices. What examples are there which suggest a different type of future for the professions, especially in terms of social purpose? Will we live in a post professional society in the future or will it be one where the professions and the society they serve work together in a different, more inclusive, democratic way for the mutual benefit of the majority? Will governments step in to make the professions provide purchasable community services? If so, what is a profession anymore? Perhaps they will become occupations with externally regulated standard performance delivery.

Alternatives require both new ways of thinking and new ways of practising. And both these new ways are available to us, if we look in the right places and stop saying "yes, but". Finland, for example, have almost solved their homelessness problem by providing people with homes (Sander, 2018). Known as "Housing First Europe" this example shows how to re-frame our thinking about the solutions available to social problems. The problem is people don't have anywhere to live; the solution is to provide them with somewhere to live, then work on the other issues which may or may not be connected. Said another way, provide people with what they need in terms of basic human rights, not what we think they need. In Australia, for example, the cashless welfare card, predominantly given to Aboriginal and Torres Strait Islander people, constrains what poor people can buy and where, and does not solve the problem of poverty.

One way of theorising this reframing and practising towards preferred and wiser futures is via the fifth wave of public health thinking (Lyon, 2003):

> A key part of this approach may be to redirect some of the energy spent on utopian schemes into accepting that humans are not perfect, to accept more of the 'messiness' which this creates and to build our future from there... using our capacity to create forms which allow us to care for and be compassionate towards others in all our imperfection... This implies moving ... towards a participative model, which empowers the individual and community and is characterised by the giving and receiving of support. (pp. 23-24)

This could be further conceptualised as civic professionalism (Sullivan, 2004) where people practise for the common good, acting to transform society and eliminate suffering borne from inequity and injustice (Frost & Hoggett, 2008). And for many individuals doing this sort of work is precisely why they worked hard to become qualified, competent professionals in the first place. Sullivan (2004) further claims that what has been missing in the professions in recent years is "action in which the professions take public leadership in solving perceived public problems" (p. 18). We would add that re-claiming the collective nature of the professions and using that collectivism will work against yet another individualised project, at risk of being co-opted by market forces.

Following Arendt this can be achieved through an "active" professional life. Kreber (2016) argues that "the changing conditions in the public sector have led to professional life increasingly taking on the forms of labour and work, at the expense of action" (p. 125), and it is this very notion of action which would enable the professions to move towards re-envisaging and re-claiming their central role in contributing to a just, fair and "good" society. This can be realised through deliberative democracy when professionals and lay people collectively, as different but equal players, work together to solve the crisis of our times. To enact this, professionals need to practise being political (as in working to change society), being visible and holding a civic set of values as central to their work: that is, they work with others to collectively re-imagine and activate a more just and equitable society. What is required in terms of professional practice then is a narrative, future-oriented imagination where we try to imagine others' point of view, recognising that there are a plurality of right views, right ways and right actions.

One of the concrete practices which flows from this is the creation of spaces for public deliberation of *different but equal* players about issues that affect people. It means having to justify one's actions and decisions in these public spaces, not just to those within the profession/s, or politicians, or the practice arena. This means recognising and embracing plurality and messiness and it takes "courage and imagination" (Kreber, 2016, p. 134). This move, we believe, could actually strengthen rather than weaken the professions as they become more embedded in, responsive to and valued by the people they purport to serve. For example:

In Porto Alegro, Brazil, local "citizen schools" are part of a move to "thick democracy" where a deliberate process of active and public participation in decision making, including the allocation of resources, takes place. As a result, the local governing body, the Popular Administration, has re-allocated resources to the most impoverished schools in addition to doubling the actual number of schools in the area (Gandin & Apple, 2002).

Balkrishna Doshi, award-winning architect, "has called on his profession to rethink the way it approaches building for the most impoverished communities… Doshi was awarded the Pritzker prize last week, in large part for the Aranya low-cost housing project. It accommodates 80,000 people with houses and courtyards linked by a maze of pathways in the city of Indore… Doshi said that architects and urban planners involved in low-income housing projects – as well as architectural education – needed to move away from their focus on the designer as individual to being far more collaborative, compassionate and invested in the dignity of those they house" (Beaumont, 2018, n.p.).

CONCLUSION

We are arguing in favour of resuming a commitment to the public good while at the same time developing a new approach to preparing for and pursuing the social purpose of professional education and practice. To achieve this, it is necessary to critically question the increasingly accepted nexus between education, industry and the professions and to challenge the taken-for-granted assumption that what is good for industry is good for most people and the planet. This we believe is one way of pursuing a preferable future, a way which embraces complexity, relationships and human messiness whilst at the same time harnessing, for the common good, the collective power already existing in the professions and made manifest in practice.

NOTES

[1] https://vuws.westernsydney.edu.au/bbcswebdav/courses/101569_2016_aut/101569%20Sustainable%20Futures%20-%20Learning%20Guide%20AUT%202016.pdf
[2] Thanks to Andy Horsfall for providing this example from his current teaching of *Sustainable Futures*, 2018.
[3] Italicised quotes are data quotes which (being unpublished) do not have page numbers listed.
[4] Welfare economics and ecological economics both strive to include a wide range of social values in the economy. The current push for indicators of genuine social progress rather than monetary progress comes from these sorts of economists.
[5] An interesting exception to this trend is provided by stimulating programs being introduced by some universities, such as the University of NSW, seeking to address major global challenges.

REFERENCES

Apple, M. W. (2006). Understanding and interrupting neoliberalism and neo-conservatism in education. *Pedagogies: An International Journal, 1*(1), 21-26.

Beaumont, P. (2018, March 12). Low-cost housing needs dignity, says Indian architect Balkrishna Doshi. *The Guardian* (online). Retrieved from https://www.theguardian.com/global-development/2018/mar/12/low-cost-housing-needs-dignity-indian-architect-balkrishna-doshi

Beck, J., & Young, M. F. D. (2005). The assault on the professions and the restructuring of academic and professional identities: A Bernsteinian analysis. *British Journal of Sociology of Education, 26*(2), 183-197.

East, L., Stokes, R., & Walker, M. (2014). Universities, the public good and professional education in the UK. *Studies in Higher Education, 39*(9), 1617-1633.

Fook, J. (1999). Deconstructing and reconstructing professional expertise. In B. Fawcett, B. Featherstone, J. Fook, A. Rossiter, & A. Rossiter (Eds.), *Practice and research in social work: Postmodern feminist perspectives* (pp. 105-120). New York, NY: Routledge.

Frost, L., & Hoggett, P. (2008). Human agency and social suffering. *Critical Social Policy, 28*(4) 438-60.

Gandin, L. A., & Apple, M. W. (2002). Thin versus thick democracy in education: Porto alegre and the creation of alternatives to neo-liberalism. *International Studies in Sociology of Education, 12*(2), 99-116.

Hazelkorn, E., & Gibson A. (2017). *Public goods and public policy: What is public good, and who and what decides?* (Working Paper No. 18, Centre for Global Higher Education Working Paper Series). London, England: Centre for Global Higher Education.

Head, T. (2017, December 17). Which countries provide free education at a university level? *The South African News*. Retrieved from https://www.thesouthafrican.com/countries-with-free-education-for-university/

Hessel, S. (2011). *Time for outrage: Indignez-vous!* New York, NY: Hachette Books.

Inayatullah, S. (2008). Six pillars: Futures thinking for transforming. *Foresight, 10*(1), 4-21.

Kreber, C. (2016). The 'civic-minded' professional? An exploration through Hannah Arendt's 'vita activa'. *Educational Philosophy and Theory, 48*(2), 123-137.

Lyon, A. (2003). *The fifth wave*. Edinburgh, Scotland: Scottish Council Foundation.

Lyons, K., & Hill, R. (2018, February 5). Vice-chancellors' salaries are just a symptom of what's wrong with universities. *The Conversation*. Retrieved from http://theconversation.com/vice-chancellors-salaries-are-just-a-symptom-of-whats-wrong-with-universities-90999

Martimianakis, M. A., Maniate, J. M., & Hodges, B. D. (2009). Sociological interpretations of professionalism. *Medical Education, 43*, 829-837.

Moreton-Robinson, A. (2000). *Talkin' up to the white woman: Aboriginal women and feminism*. Brisbane, Australia: University of Queensland Press.

Olssen, M., & Peters, M. A. (2005). Neoliberalism, higher education and the knowledge economy: From the free market to knowledge capitalism. *Journal of Education Policy, 20*(3), 313-345.

Power, S. (2002). *A problem from hell: America and the age of genocide*. New York, NY: Basic Books.

Rosenberg, J. P., Horsfall, D., Leonard, R., & Noonan, K. (2017). Informal care networks' views of palliative care services: Help or hindrance? *Death Studies, 42*(6), 362-370.

Rossiter, A. (1996). Finding meaning for social work in transitional times: Reflections of change. In N. Gould & I. Taylor (Eds.), *Reflective learning for social work: Research, theory and practice* (pp. 141-151). England: Ashgate Publishing.

Sander, G. (2018, March 21). Finland's homeless crisis nearly solved: How? By giving homes to all who need. *The Christian Science Monitor*. Retrieved from https://www.csmonitor.com/World/Europe/2018/0321/Finland-s-homeless-crisis-nearly-solved.-How-By-giving-homes-to-all-who-need

Sullivan, W. M. (2004). Can professionalism still be a viable ethic? *The Good Society, 13*(1), 15-20.

Thornton, M. (2015, November 4). A focus on private investment means universities can't fulfil their public role. *The Conversation*. Retrieved from https://theconversation.com/a-focus-on-private-investment-means-universities-cant-fulfil-their-public-role-45094

Universities Australia. (2013). *A smarter Australia: An agenda for Australian higher education 2013-2016*. Canberra, Australia: Author.

Wilensky, H. L. (1964). The professionalization of everyone? *American Journal of Sociology, 70*, 137-158.

Debbie Horsfall PhD (ORCID: https://orcid.org/0000-0002-9266-6234)
Western Sydney University, Australia

Joy Higgs AM, PhD (ORCID: https://orcid.org/0000-0002-8545-1016)
Emeritus Professor, Charles Sturt University, Australia
Director, Education, Practice and Employability Network, Australia

STEVEN CORK

7. OUR PLACE IN SOCIETY AND THE ENVIRONMENT

Opportunities and Responsibilities for Professional Practice Futures

RESPONSIBILITIES OF THE PROFESSIONS AND PROFESSIONALS

The professions emerged as institutions that were trusted and relied upon to help society deal with complex concepts and information that required training and experience beyond what most people could achieve (Susskind & Susskind, 2015). As discussed by Cork and Alford in Chapter 3 of this volume, questions have been raised about whether the responsibilities embodied in this concept of the professions are being met. Cork and Alford consider a range of plausible futures in which, at one extreme, institutions resembling today's professions might continue to play the social roles they now play or, at another extreme, such roles might be discharged by very different institutions. Similarly, the types of people with specialist knowledge and experience that we currently call professionals (using the term to mean members of professions) might still exist in the future or might be replaced to a greater or lesser extent by combinations of artificial intelligence and humans with broad skills in managing and synthesising diverse information and ideas.

In this chapter I argue that, in the future, the successors of today's professions and professionals face at least one major challenge and one major opportunity. The challenge will be to regain the trusted position that the professions once had. The opportunity will be to play a pivotal role in helping society gather, analyse and apply information, across all disciplines and "ways of knowing the world", so that humans can both envision and achieve sustainable futures, noting that humanity currently lacks a clear vision for what the words *sustainable futures* mean in practice.

GUIDANCE FOR THE PROFESSIONS AND PROFESSIONAL PRACTICE

If the successors of today's professions and professionals are to have a responsible place in society and the environment, from where might they receive guidance? One answer to this question comes from the dialogue that has been emerging around the word *sustainability* over the past five decades. Since the 1970s, a global dialogue has been under way about how humans could and should interact with one another and the natural environment to ensure that our species persists, a good quality of life is maintained for all humans, and our moral and ethical responsibilities to other species are observed. The concepts of *sustainability* and *sustainable development* have been at the heart of this dialogue.

Sustainable development has been defined as "development that meets the needs of the present without compromising the ability of future generations to meet their own needs" (United Nations, 2018, n.p.). The long dialogue about how this definition could be applied began by considering metaphors like the *triple bottom line* to represent the *environmental, social* and *economic* dimensions of sustainability and has evolved into a globally agreed set of 17 Sustainable Development Goals (see Figure 7.1) covering all aspects of a decent and responsible human life, from fundamental humans needs, through aspects of health, wellbeing and prosperity, to responsible management of planetary process and peace, justice and equality (United Nations, 2018).

1.	No More Poverty
2.	Zero Hunger
3.	Good Health and Well Being
4.	Quality Education
5.	Gender Equality
6.	Clean Water and Sanitation
7.	Affordable and Clean Energy
8.	Decent Work and Economic Growth
9.	Industry, Innovation and Infrastructure
10.	Reduced Inequalities
11.	Sustainable Cities and Communities
12.	Responsible Production and Consumption
13.	Climate Action
14.	Life Below Water
15.	Life on Land
16.	Peace, Justice and Strong Institutions
17.	Partnerships for the Goals

Figure 7.1. The 17 Sustainable Development Goals agreed through the United Nations (United Nations, 2018).[1]

Several books could be (and have been) written about how these goals have been developed and the challenges of implementing them. In this chapter I want to focus on two key challenges that are especially relevant to future professional practice: (1) *envisioning* possible futures that combine all of these goals and maintain them indefinitely; and (2) *gathering and analysing* the sort of information required to help society consider how these goals might be integrated. If the successors of today's professions and professionals are to be socially and environmental responsible they must, as a minimum, follow the guidance that emerges from the dialogue about sustainable futures. But I argue that they can, and should, do more by taking a lead in addressing the two challenges listed above. Below I explore how not only minimal actions but also leadership roles could be performed to address these outcomes.

SUSTAINABILITY AS AN INTEGRATION OF VALUES AND ENTITIES

A wide range of approaches has been developed to identify criteria and indicators for social, environmental and economic sustainability. Fewer examples exist of efforts to consider how these three sets of values can be integrated and what an integrated sustainable future might be like. Here I will draw on a recent analysis by Boyer et al. (2016) that provides a framework for considering ways in which the relationships between social and other aspects of sustainability have been framed in academic and public discourse. I suggest that their conclusions can be generalised to the framing of relationships between social, environmental and economic aspects of sustainability more broadly (see Table 7.1).

Table 7.1. How sustainability has been framed (adapted from Boyer et al., 2016).

Frame	Examples
Value-categories[a] as *stand-alone pillars*	Many schemes for assessing sustainability list indicators for environmental, social and economic values separately with little guidance on how to deal with synergies, trade-offs or integration.
Value-categories as *constraints* on one another	It appears that the underlying imperative in many approaches to assessing business sustainability is financial performance, which is adjusted only when there seems to be unacceptable social or environmental impacts.
Value-categories as *preconditions*	Some models for considering relationships between values portray a sustainable environment as being a necessary precondition for a sustainable society and a sustainable society as a precondition for a sustainable economy, while others have the environment and society nested within the economy.
Value-categories as *causal mechanisms* of change	It is variously argued that social capital is required for moving society towards sustainability, or that a strong economy allows people to think about environmental sustainability etc.
A *fully integrated*, locally rooted and process-oriented approach to sustainability	Use of inclusive dialogue to bring diverse parties together to consider the full range of elements of community life, drawing on local knowledge and grassroots movements as a guide for policy and action, and viewing social, economic and environmental imperatives as overlapping in local experience.

[a] The three "triple bottom line" categories of value: social, environmental and economic.

Boyer et al. (2016) conclude that none of the first four frames in Table 7.1 are consistent with what we know about how humans survive and thrive in the world. They propose that the distinction between social, environmental and economic values is an historical artefact that not only perpetuates divisions between disciplines (and, I add, professions) but also excludes other forms of knowledge and understanding. They argue that the fifth, fully integrated, approach requires governance arrangements that facilitate the sort of dialogue required to seek sustainable futures. That dialogue, I argue, should be supported by collection and sharing of relevant information among stakeholders. In the context of this chapter, it is logical to see the successors of today's professions and professionals as playing key roles as institutions and actors in these governance arrangements and dialogue. That is, future professionals need to increase their roles in human systems leadership as opposed to, or in addition to, more local client-centred roles.

Henceforth in this chapter, I will review a variety of approaches to assessing progress towards sustainability and discuss how these might influence, and be influenced by, the successors of today's professions and professionals. I will refer to analyses that arise from the first four types of frames listed in Table 7.1 as *partial assessments* and to analyses arising from the fifth frame as (whole) *integrated assessments*. In the following sections I argue that partial assessment approaches will continue to guide the successors of today's professions in relation to their social, economic and environmental responsibilities but that the most important opportunity for these successors is to play a pivotal role in applying integrated assessment approaches, many of which have not yet been developed or even conceived. A key part of their role, then, is to imagine and develop such integrated approaches.

PARTIAL ASSESSMENTS AND THE SUCCESSORS OF TODAY'S PROFESSIONS

Today, businesses, the professions and professionals within them mainly seek to discharge their social and environmental responsibilities by limiting their activities when those activities have undesirable effects on society or the environment (i.e. they take a *constraint* approach). Other examples of constraint approaches are the many metrics that assess the effects of human activities on the environment and attempt to identify acceptable limits to the effects. For example, Rockström et al. (2009) proposed a set of measures that define the biophysical limits, or a *safe operating space*, within which the planet is likely to continue to maintain human life. Early efforts have also been made to explore a *social safe operating space* (Alford et al., 2012) but these are not well advanced. The United Nations' Sustainable Development Goals are an attempt to define ethical and equitable boundaries for environmental, social and economic values (United Nations, 2018). For example, they challenge us to define terms like *quality, equality, inequalities, decent* and *responsible* in relation to several goals, while calling for *no* poverty and *zero* hunger.

State of the Environment (SoE) reporting in many countries considers drivers of change, pressures on the environment, measures of the state (condition) of the environment, the implications of changes to state, and the adequacy of responses to these changes and their implications (the DPSIR model).[2] Most SoE assessments, however, struggle to draw clear conclusions about the implications for humans, beyond obvious effects of air, land and water pollution. In SoE reporting in the past decade and a half, concepts like *Ecological Footprint, Ecosystem Services* and *Resilience* have been drawn on to consider implications. The latter two concepts will be discussed later in this chapter. Here I will consider Ecological Footprint as an example of a metric that can guide, and has guided, the ways in which professions and professionals discharge their social and environmental responsibilities.

Ecological Footprint is a metaphor that has been employed to help people make sense of their complex relationship with the environment (Global Footprint Network, 2018). This and other approaches that deal primarily with the environment pillar of sustainability have been major vehicles for businesses and communities to assess and demonstrate their contributions to sustainable futures. It has been argued that such approaches focus primarily on the environment pillar of sustainability and, therefore, are of the *single pillar* and/or *constraint* type of frame described by Boyer et al. (2016) (see Table 7.1). It could also be argued, however, that dissecting the details of footprint analysis yields many insights into social and economic aspects of sustainability and even goes some way towards integrating the three sustainability pillars (i.e. social, environmental and economic).

Ecological Footprint accounting measures what humans demand from the environment and calculates what types and amounts of land are needed to supply goods and services to meet these demands. Examples of demands include: plant-based food and fibre products; livestock and fish products; timber and other forest products; and space for urban infrastructure. There is also a need for land to absorb wastes produced directly by humans and indirectly by the various manufacturing, processing and other activities that accompany human lives (including carbon emissions associated with all of these processes, often termed the *Carbon Footprint*). Ecological Footprints can be calculated at a range of scales from the basis of an individual or household to professions, companies, industries, nations and the globe.

For example, a recent study surveyed self-selected (and therefore not necessarily representative) members of three groups of professions: conservation-related professions; economics; and medical sciences. It estimated their per capita carbon footprints from the following information: mode of travel to work; flights per year; energy-saving measures at home; offsetting emissions generated by energy use, travel, recycling, composting, etc.; generation of food waste; consumption of meat or fish; use of bottled water; number of children (had or hope to have); and ownership of cats and dogs (Balmford et al., 2011). The environmentalists had lower carbon footprints but the differences were small.

Information provided by metrics like Ecological Footprint can be, and is being, used to encourage professions and individual companies to manage their footprint and to allow them to assess their contributions to society's environmental goals (Murray, 2012). The sophistication and accuracy of such measures is likely to increase in many plausible futures, due to the combination of more powerful technologies for gathering and analysing information together with greater integration across disciplines and professions (see Chapter 3).

INTEGRATED ASSESSMENTS AND THE SUCCESSORS OF PROFESSIONS

In this section I focus on two examples of approaches to integrating the components of sustainability: one that tries to build links between the environmental, social and economic pillars and another that blurs the distinctions between the three pillars. The concept of *ecosystem services* attempts to explain, in everyday language, how the environment supports human wellbeing in both social and economic terms (see Figure 7.2).

A typical definition of ecosystem services is: "the direct and indirect contributions of ecosystems to human well-being" (Sukhdev et al., 2010). Although this definition suggests a *precondition* frame (as introduced in Table 7.1), the consideration of multiple means of feedback and influences among ecosystems, social values and drivers of change make this approach, I suggest, more integrative.

Refinement of the concept of ecosystem services has seen a sometimes uneasy coming together of ideas from social, environmental and economic disciplines. It is rooted in the idea of services as emerging from the transformation of resources in ways that create benefits to humans. Ecosystems are complex interactions among living and non-living components of the environment (e.g. forests, grasslands, riverine ecosystems, marine ecosystems) and these interactions mediate major transformations of resources, many of which rival or exceed what can be achieved cost-effectively by engineering or other interventions by humans (e.g. maintenance of atmospheric gases, large-scale filtration and purification of water, or widespread control of potential pest species).

The concept of ecosystem services (see Figure 7.2) aims to name and categorise the benefits coming to humans from nature and to find ways to estimate their contributions in both economic and social terms. The tools for economic analysis mostly existed previously (Sukhdev et al., 2010) but ecosystem services approaches sought to engage more people and more types of knowledge in dialogue about what aspects of the environment had "worth" (be it monetary or other worth) (Cork et al., 2012). In this way, the ecosystem services approaches introduced more "ways of knowing" than economics or ecological sciences.

Crucially, however, it became apparent that linking the environment to social values required more thinking about fundamental human needs. In the illustration of ecosystem services approaches (see Figure 7.2) note the arrows indicating "strategies and interventions". These are powerful pathways of influence in coupled social-ecological systems. As explained later in this chapter, the successors of today's professions and professionals should be able to influence such mediation if they build their capacity to deal with the complexities of such systems. This will require them to not only take advantage of emerging integrative technologies but also enhance cooperation across disciplines to achieve transdisciplinarity.

Boyer et al. (2016) provide several other examples of integrative approaches to sustainability. For example, they refer to the work of a specialist in tribal community development who "has worked with several colleagues to develop a guide for Indigenous community self-assessment that moves communities beyond discussions of economic development to a perspective that includes culture, identity, history, and other key elements of community lives" (p. 12).

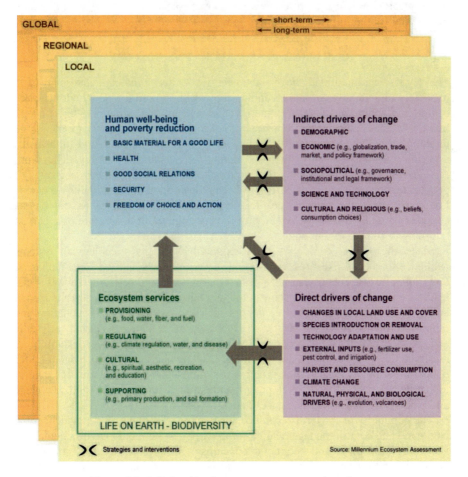

Figure 7.2. Relationships between ecosystem services, human wellbeing and drivers of change (by Philippe Rekacewicz, Emmanuelle Bournay, UNEP/GRID-Arendal).[3]

They point out that this approach of inclusive dialogue, while still relatively uncommon, is not new, having been proposed for several decades as a vital approach for addressing societies' "wicked problems" (Australian Public Service Commission, 2007; Rittel & Webber, 1973). What often prevents this approach from being applied is the absence of institutions and situations that allow inclusive dialogue to occur (Cork et al., 2015; Walker et al., 2009). I argue later that the future incarnations of the professions can play a role in facilitating such dialogue that is vital for the future of humanity.

RESILIENCE AS AN INTEGRATIVE CONCEPT

The key issues highlighted above – the need for inclusive dialogue across all parts of society and all ways of understanding the world – is reinforced by another area of thinking about the future of humanity: the concept of *Resilience*. This concept, like Ecological Footprint and Ecosystem Services, emerged as a way to encourage thinking about how humanity might survive future shocks and upheavals, especially in the interim while we search for sustainable futures. The word *resilience* is used across society to describe the ability of individuals or groups of people to cope with stress or hardship. It has the connotation of "bouncing back". For some time, this same word has been used in disciplines as diverse as engineering and psychology, again with the connotation of a structure or a person returning to some previous state.

Since the 1970s, ecologists have explored what gives ecological systems resilience to shocks like fires, flooding, disease outbreaks and the like (Walker & Salt, 2006, 2012). This research suggested that much could be learned about social resilience from ecological systems; and a major international research agenda emerged, funded by philanthropic donors such as the Rockefeller Foundation and others (Resilience Alliance, 2011). The inclusion of resilience as a *goal* for public and private policies and programs has become widespread. For example, in Australia the word *resilience* appears in the heading and/or key objectives of major plans for nationally critical infrastructure, risk management across a wide range of industries, early childhood development, and national strategies for climate change and biodiversity management (Cork, 2010).

The popularity of the term *resilience* appears to come in part from evidence that command and control management of complex issues – not only environmental but also social, business and economic concerns – has frequently had unintended negative consequences. This has led many government, business and community leaders to conclude that building capacity to cope with shocks might be a better strategy in many cases than applying "least worst" options or interventions based on minimal understanding of the dynamics of complex problems.

This mode of thinking is still far from widespread but it is consistent with analyses of how to tackle "wicked problems" in public policy. A landmark report to the Australian Public Service in 2007 (Australian Public Service Commission, 2007) identified three broad options for dealing with complex, poorly defined and contested (i.e. wicked) issues: *Authoritative* strategies (giving some group authority to devise the solution); *Competitive* strategies (e.g. allowing markets, advocacy or other competitive processes to decide whose ideas and preferences prevail); and *Collaborative* strategies (facilitating exchange of ideas to more clearly define the issues, identify areas of agreement and disagreement after alternative viewpoints are understood, and seeking partnerships, joint ventures and the like).

Strategies allowing the free exchange of ideas and fluid governance arrangements are consistent with a core insight from the resilience research: that allowing systems to self-organise leads to greater resilience than imposing organisation, especially when that organisation involves control from one or a few central points. In ecological systems, self-organisation takes the form of multiple species applying diverse strategies, alone and in synergy with others, to cope with changing environments. Successful solutions emerging after processes of experimentation and "trial and error" (admittedly, not a competition-free process but one that allows innovation and cooperation). In social systems, self-organisation requires polycentric governance that allows experimentation at all levels of societal hierarchies (Andersson & Ostrom, 2008). There is increasing evidence that imposing top-down constraints on social processes, especially ones aimed at reducing redundancy and diversity in the system, frequently decreases resilience and increases vulnerability to shocks.[4]

Some other key insights from research on resilient ecological and social systems (Cork, 2010; Resilience Alliance, 2011; Walker & Salt, 2006, 2012) include:

- resilience cannot be only about returning to a previous condition as resilient ecosystems, social systems and individuals almost always undergo some change to allow them to adapt to changed circumstances
- resilience emerges from complex interactions between many parts of systems and so understanding resilience requires understanding whatever system we are interested in
- if we set resilience as a goal for a system, we must be clear about what system we are talking about, what aspects of that system we want to have resilience and what challenges we want them to be resilient to (which challenges all of society to be clear about our values and expectations)
- sometimes investing in some types of resilience can result in underinvestment in others
- resilience is not always desirable (for example, some despotic political regimes or dysfunctional social-support systems can be hard to change because the feedbacks within the system keep it operating in an undesirable way)
- resilience in social systems requires inclusive leadership, capacity to learn and adapt, commitment to considering and watching for early warning signs of multiple plausible futures and preparing for not only possible challenges but also opportunities.

Methods for assessing resilience of coupled social-ecological systems (and also the related concepts of adaptability and transformability) are under development (Resilience Alliance, 2011). They require expertise and resources for systems analysis and engagement of all relevant stakeholders. This is the sort of role that the successors of today's professions and professionals could, and should, play.

CONCLUSION

We cannot be sure what institutions or individuals will meet society's needs for making sense of complex information and issues in the future, or how they will do that. At one extreme we might see the necessary skills developed by individuals in a similar way to what we see today (but presumably with many more tools to collect, manage and share information) and the practices of those individuals might still be coordinated and regulated by professions. At the other extreme we might see more provision, analysis and even interpretation of information by artificial intelligence, severe blurring of boundaries between disciplines and the disappearance of anything resembling today's professions (see Chapter 3).

Assuming successors to professions and professionals exist in the future, they might simply conform to rules identified by society that define how much adverse impact "professional" practices can have. This would be a *constraint* approach to sustainability (see Table 7.1).

Such an approach would not see the successors of the professions regaining their status as trusted *leaders* of society. A more likely scenario, in my view (or maybe I am just being hopeful), is that people with exceptional training, skills and ability in dealing with complex issues and using advanced artificial intelligence as a tool will work together in some sort of formal or informal institution to help society achieve the integration across all sorts of knowledge and value systems.

These institutions might not be called *professions*, but they would serve a similar role to that of the earliest professions – occupying a position of trust in helping society deal with complexity that is beyond the ability of ordinary members to engage with. Virtually all published thinking about futures of work (see Chapter 3) agrees that this advanced capacity for integrative thinking and action will be vital in most plausible futures.

Some might argue (and have argued) that artificial intelligence can do at least as good a job as humans of acting in ethical, moral and trustworthy ways. Be this as it may, at least in the next few decades, while we wait to see what the potential of artificial intelligence is, there are likely to be two intersecting opportunities for the successors of today's professions and professionals: (1) unprecedented demand for their help and leadership from society in dealing with complex, wicked problems; and (2) the need for them to develop the equally unprecedented ability to meet that demand as information, communication and other technologies go through exponential increases in capabilities.

NOTES

[1] https://www.un.org/sustainabledevelopment/sustainable-development-goals/
[2] DPSIR (Drivers, Pressures, State, Impact, Response model of intervention) is a casual framework for describing the interdependent interactions between society and the environment.

[3] Available from https://www.grida.no/resources/6059
[4] See numerous case studies at http://www.regimeshifts.org/

REFERENCES

Alford, K., Manderson, L., Boschetti, F., Davies, J., Hatfield-Dodds, S., Lowe, I., & Perez, P. (2012). Social perspectives on sustainability and equity in Australia. In M. R. Raupach, A. J. McMichael, J. J. Finnigan, L. Manderson, & B. H. Walker (Eds.), *Negotiating our future: Living scenarios for Australia to 2050* (Vol. 1, pp. 93-114). Canberra, Australia: Australian Academy of Science.

Andersson, K. P., & Ostrom, E. (2008). Analyzing decentralized resource regimes from a polycentric perspective. *Policy Sciences*, *41*(1), 71-93.

Australian Public Service Commission. (2007). *Tackling wicked problems: A public policy perspective.* Canberra, Australia: Commonwealth of Australia.

Balmford, A., Fisher, B., Green, R. E., Naidoo, R., Strassburg, B., Turner, R. K., & Rodrigues, A. S. L. (2011). Bringing ecosystem services into the real world: An operational framework for assessing the economic consequences of losing wild nature. *Environmental and Resource Economics*, *48*(2), 161-175.

Boyer, R., Peterson, N., Arora, P., & Caldwell, K. (2016). Five approaches to social sustainability and an integrated way forward. *Sustainability*, *8*(9), 878-818.

Cork, S. (2010). Resilience of social–ecological systems. In S. Cork (Ed.), *Resilience and transformation: Preparing Australia for uncertain futures* (pp. 131-142). Melbourne, Australia: CSIRO Publishing.

Cork, S., Gorrie, G., Ampt, P., Maynard, S., Rowland, P., Oliphant, R., Reeder, L., & Stephens, L. (2012). *Discussion paper on ecosystem services for the Department of Agriculture, Fisheries and Forestry* (Final Report). Retrieved from http://www.agriculture.gov.au/ag-farm-food/natural-resources/ecosystem-services/ecosystem-services-report

Cork, S., Grigg, N., Alford, K., Finnigan, J., Fulton, B., & Raupach, M. (2015). *Australia 2050: Structuring conversations about our future*. Canberra, Australia: Australian Academy of Science. Retrieved from https://www.science.org.au/files/userfiles/support/reports-and-plans/2015/australia-2050-vol-3.pdf

Global Footprint Network. (2018). *Ecological footprint*. Retrieved from https://www.footprintnetwork.org/our-work/ecological-footprint/

Millennium Ecosystem Assessment. (2003). *Ecosystems and human well-being: A framework for assessment*. Washington, DC: Island Press.

Murray, S. (2012, April 24). Companies try to reduce humanity's footprint. *Financial Times*. Retrieved from https://www.ft.com/content/99a6b83c-87ce-11e1-ade2-00144feab49a

Resilience Alliance. (2011). *Resilience: 40 years of resilience research and thinking*. Retrieved from http://www.resalliance.org

Rittel, H. W. J., & Webber, M. M. (1973). Dilemmas in a general theory of planning. *Policy Sciences*, *4*(2), 155-169.

Rockström, J., Steffen, W., Noone, K., Persson, A., Chapin, F. S., Lambin, E. F., Lenton, T. M., Scheffer, M., Folke, C., Schellnhuber, H. J., Nykvist, B., de Wit, C. A., Hughes, T., van der Leeuw, S., Rodhe, H., Sörlin, S., Snyder, P. K., Costanza, R., Svedlin, U., Falkenmark, M., Karlgerg, L., Corell, R. W., Fabry, V. J., Hansen, J., Walker, B., Liverman, D., Richardson, K., Crutzen, P., & Foley, J. A. (2009). A safe operating space for humanity. *Nature*, *461*(7263), 472-475.

Sukhdev, P., Wittmer, H., Schröter-Schlaack, C., Nesshöver, C., Bishop, J., ten Brink, P., Gundimeda, H., Kumar, P., & Simmons, B. (2010). *The economics of ecosystems and biodiversity: Mainstreaming the economics of nature: A synthesis of the approach, conclusions and recommendations of TEEB*. European Communities.

Susskind, R., & Susskind, D. (2015). *The future of the professions: How technology will transform the work of human experts*. Oxford, England: Oxford University Press.

United Nations. (2018). *The sustainable development agenda*. Retrieved from https://www.un.org/sustainabledevelopment/development-agenda/
Walker, B., & Salt, D. (2006). *Resilience thinking: Sustaining ecosystems and people in a changing world*. Washington, DC: Island Press.
Walker, B., & Salt, D. (2012). *Resilience practice*. Washington, DC: Island Press.
Walker, B., Barrett, S., Polasky, S., Galaz, V., Folke, C., Engström, G., Ackerman, F., Arrow, K., Carpenter, S., Chopra, K., Daily, G., Ehrlich, P., Hughes, T., Kautsky, N., Levin, S., Mäler, K. G., Shogren, J., Vincent, J., Xepapadeas, T., & de Zeeuw, A. (2009). Environment. Looming global-scale failures and missing institutions. *Science, 325*(5946), 1345-1346.

Steven Cork PhD (ORCID: https://orcid.org/0000-0002-3270-4585)
Crawford School of Public Policy
Australian National University, Australia
Ecoinsights, Australia
Australia21, Australia

SANDY O'SULLIVAN

8. PRACTICE FUTURES FOR INDIGENOUS AGENCY

Our Gaps, Our Leaps

For many First Nations'[1] communities, self-determination and community-led approaches are crucial to building capacity, enhancing agency and in maintaining control over our future. In writing this chapter I support the principle that it is an essential resetting of Indigenous Peoples role as "subjects" of research, that we ensure that – as academic writers – we do not distance ourselves from our own subjectivity, by tacitly aligning with our communities (Wiradjuri, named nations) and meta-communities (Indigenous, Aboriginal, First Nations) by adopting "our" and "we" to describe effect, in place of "they" and "them".

As a major site of colonial invasion, Australia has struggled with goals and actions aimed at supporting Indigenous aspirations and agency (Nakata, 2010; Pascoe, 2016). The most sympathetic reading of the colonial project would be that its goal was to ensure that assimilation and dominance was complete with the colonised (people) only being sustained if we followed the colonial path of least resistance. As with other First Nations' Peoples the policies that enacted colonisation have led to successive generations of poorer social determinants for our peoples, than for our non-Indigenous counterparts (Fforde et al., 2013; Pascoe, 2018).

In 2007, the Council of Australian Governments committed to "closing the gap on Indigenous disadvantage", with yearly reporting provided by the Australian Government (Australian Human Rights Commission, 2008). A 10-year anniversary assessment of the policy was released by the Australian Human Rights Commission, a key agency in proposing the initiative, and one that has carriage of the ongoing Close the Gap campaign. The main criticism, beyond the failure to meet set targets, was that the government policy responses lacked any meaningful evaluation, suggesting that the measures represented a "failure of accountability and good governance by the Federal Government" (Australian Human Rights Commission, 2018, p. 8). The report also offered a criticism that parity is measured only within a deficit space, and government efforts were "…heavily skewed toward the costs of reacting to the outcomes of disadvantage rather than investments to reduce or overcome disadvantage" (ibid, p. 8).

The difficulty with the deficit space located in the policies around Closing the Gap is that in measuring only parity, these measures cannot speak to aspiration, and that the markers as equivalence cannot measure excellence and do not accommodate intentional difference (Moore, 2012). While the gap reflects disadvantage and dispossession within the colonial project and an urgency remains to ensure that education, health and social disadvantage are addressed, there are a number of Indigenous-led initiatives that propose a strengths-based approach to changing both the narrative and the outcomes across our communities (Fredericks, 2013; McKinnon, 2014; Janke, 2009). These ideas will support us in not only closing the gaps that exist by developing and using our own culturally appropriate and endorsed strategies, but by leaping across these gaps to demonstrate the unquestionable reality of the innovation and strength that have sustained our cultures and communities for more than 60,000 years.

Historically, education has been a colonial tool to manage our compliance and punish the retention of our traditional and cultural Knowledges (Martin, 2003; Price, 2015). In order to recalibrate this relationship, many First Nations' scholars are working towards a mainstream acknowledgement of the value of our cultural Knowledges (Behrendt et al., 2012; Bond, 2014; Kovach, 2009), albeit often with the burden firmly placed on those of us located at the site of (academic) engagement. It does, however, provide those of us within the system, with the opportunity to challenge the destructive tenets of the colonial academy (Bodkin-Andrews & Carlson, 2016; Bodkin-Andrews & Craven, 2013).

This chapter focuses on some of the inroads to change that are being made in higher education in order for us to shape our futures and create an imaginary world? that supports aspiration, agency and participation across all aspects of our lives. Changes to Indigenous inclusion in higher education have been led by our own academics. Alongside our allies within the system, these new approaches encourage the voices from our communities, understanding that students can, and should, be participating across the academy in every way (including academic governance as well as learning and course design), and by reshaping and replacing the tools of colonisation to tell our national story in education, aspiration and self-determination.

OUR *EDUMACATION*, OUR PARTICIPATION

I recognise that as a colonised Aboriginal woman working in the academy my journey has been both one of compliance and resistance. Like many of my Indigenous colleagues, I left school young – at 13 – and went back to education much later, with both a fear of the system and an excitement for what it may offer (O'Sullivan, 2015). I have spent nearly 30 years in higher education, have succeeded in becoming one of the too-few Aboriginal people with a PhD – and the even fewer who have reached the professoriate. For those of us who have achieved what may be seen as academic success, we participate in spite of the unwritten rules of alterity (or otherness) in which Indigenous academics operate (Asmar & Page, 2009; Bodkin-Andrews & Carlson, 2016; Bond, 2014).

The term "edumacation" has been borrowed from African-American English, where it operates as an infixer to form an "ironic pseudo-sophistication" (Yu, 2004). When deployed in Aboriginal English, the word has similar use, where it both diminishes the value of the formal education system as a sole remedy to knowledge attainment while proposing a more comprehensive understanding of what education can and should be. When Aunty Ruby Langford Ginibi deployed the term across her series of talks about her own engagement with education and her understanding of the colonial system, she implied a scaffolding of her knowledge where formal learning met cultural and individual experience forming a more complete idea than the term "education" suggests (Jones, 2012). It is, essentially, a strengths-based approach to a learning system that requires specific and ongoing negotiation, and it fits outside of the idea of simple credentialing of learning.

In her central thesis, critical race scholar, Audre Lorde suggested that the "M(m)aster's tools will never dismantle the master's house" (Lorde, 2018, p. 13). Maintaining our learning and research opportunities wholly within the structure and endorsement of the institution is a recipe for remaining in deficit, meeting the master's requirements (not our own goals) and failing to close the gap that the masters define (Fforde et al., 2013; Fredericks, 2009). To challenge this, First Nations' scholars at universities around the world are working within the system to bring in Elders and other community members, to alter and reshape our strategies to better serve our communities (Bodkin Andrews & Carlson, 2016). The establishment of National Aboriginal and Torres Strait Islander Education Consortium (NATSIHEC) and the World Indigenous Nations Higher Education Consortium (WINHEC) engage self-determination principles that draw from our Communities, and provide an informed position for their needs and requirements.

The steps towards substantial university reform and a response to the work being undertaken by our scholars on Indigenous self-determination, were mapped in the 2012 *Review of Higher Education Access and Outcomes for Aboriginal and Torres Strait Islander People.* The Review, commissioned by the Australian Government, proposed a range of recommendations for improving our participation and engagement, and the 35 recommendations were equally split into what universities, government and communities must do to provide greater agency and opportunities for Indigenous Peoples and Communities (Behrendt et al., 2012).

In 2017, the peak body to which all Australian universities belong, Universities Australia, responded to many of the findings of the Review. From the findings, an Indigenous Strategy was formed that the Vice-Chancellors of all Australian universities agreed to recognise (Universities Australia, 2017). The key author was Aboriginal academic leader, Leanne Holt. In addition to a consolidation of the call for embedding of Aboriginal and Torres Strait Islander Knowledges in all university curriculum it recommended a range of positive measures that would set the path for both parity in our participation and would seed greater understanding of our communities in the wider population. Each measure was positively charged and a future map was created for reaching these targets, cleverly creating little opportunity for universities to deviate from the plan and timeline. These measures included increasing our student and staff numbers, providing a greater understanding in the mainstream community of our cultures and Knowledges, and importantly, also assuring a connection back to our communities to reflect needs, aspiration and agency, along with a Vice-Chancellor-level committed timeline for these measures to be met.

From the perspective of this chapter, the most substantial finding of the Report was not the anticipated parity targets. While significant, these targets still frame the measures associated with "closing the gap". Instead, the most substantial adjustment to the status quo is the expectation that by 2028 our students would have parity across all discipline areas. This change will support agency for Indigenous Australians in participating at all levels of our society well beyond gap management. The report describes the exponential growth of Indigenous student numbers up to 7.1% between 2014 and 2015, three times higher than the non-Indigenous growth rate. Conversely – and to be addressed by the application of the 2017 Universities Australia Indigenous Strategy – our participation has been dire across specific areas, including science, technology, engineering, mathematics, medicine, accounting and business. It is no coincidence that these areas frequently bear no direct relationship to any concept or achievability of "closing the gap".

Indigenous standpoint theory posits that meaningful cultural inculcation radiates from a central base of understanding within the culture and is more difficult to share, than to know (Nakata, 1998). Even a rudimentary understanding of cultural experience suggests that learning will fail to embed in a meaningful way if it is not connected to other meaningful experiences (Bourdieu, 2013). Will these changes result in non-Indigenous Australians and international students engaging more with our communities and a greater understanding of our culture? Pressure from within and without will only go so far, but the process of bringing our non-Indigenous counterparts – in the form of academic colleagues – in on these positively charged compliance measures in the way that the Strategy suggests, could support both the engagement and the dismantling of structures from which we have been excluded, and a greater understanding of who we are, in the general population.

OUR AGENCY, OUR SPECULATION

P'urhepecha scholar Michael Lerma's treatise on sovereignty and agency for First Nations across the United States, argues that a current "Indigenous Spring" has seen nations reinventing themselves as agile and capable of deploying change across their societies in contrast to the behemoth of major nation states wavering within a post-Fordian environment (Lerma, 2014). While the colonial frames that contain the US capacity for this change agility varies somewhat from the Australian experience, what we share with our international counterparts is an opportunity for aspiration that we control our own destinies (Evans & Williamson, 2017).

For nearly a decade, Aboriginal leadership scholar Michelle Evans has been running the Murra Indigenous Business Master Class with the Melbourne Business School. A fully funded professional course for emerging and established Aboriginal and Torres Strait Islander business owners, the program supports a level of networking and knowledge attainment that is usually the reserve of longer-form MBA programs. In recognising that there is a lot of ground to make up, this shorter course targets those already engaged in business who require the connection and systematic support provided to participants in a mainstream business program

(Evans & Williamson, 2017) and inherent in the colonial system of power (Nakata, 2010).

Essential to the success of these strengths-based approaches is that we – as First Nations' Peoples – must be supported in defining our own aspirations, developing our own approaches and measuring our own success. The Australian Government Indigenous Procurement Policy requires that businesses involved in schemes driven by this policy must be more than 51% Indigenous controlled, and it is through the work of the Indigenous-run *Supply Nation* that a major database of verified Indigenous businesses is endorsed. Their role and the role of leaders across this rapidly increasing area of business growth, are centred on ensuring that there is "nothing about us, without us" (Heckenberg et al., 2016). This initiative relies on both the pressure of leaders, like Indigenous lawyer – and academic – Terri Janke, whose work has leveraged claims from First Nations' Communities relating to cultural rights and Knowledges (Janke & Sentina, 2018), and through the increasing number of people in our communities who are taking on leadership roles in business.

OUR ASPIRATIONS "FIRST THINGS FIRST"

If aspiration and self-determination must come from within a cultural group, imagine the outcry and resistance if the United States defined and framed the capacity and growth for Russians? For most First Nations' Communities around the world denial of self-determination has been a central tool of the coloniser (Lerma, 2014; Kovach, 2009; Moreton-Robinson, 2011). Importantly, and recalling Lorde (2018), externally imposed values cannot assess challenges to the mainstream goals, nor recognise growth made possible through dissent.

In 2017, a group of Aboriginal and Torres Strait Islander people were brought together to form a Referendum Council to discuss the recognition of our Peoples in the Australian Constitution. Rather than the anticipated statement on constitutional recognition expected by the Australian Government, the forum delivered the *Uluru Statement from the Heart* (Referendum Council, 2017). The Statement called for reform over recognition, and "...for the establishment of a First Nations Voice enshrined in the Constitution" (n.p.). The gathering and findings that had been commissioned by the Australian Government were then summarily dismissed by the Australian Prime Minister as impractical, indicating that no parts of it would be considered (Langton, 2018).

As a response to the Statement and the dismissal, *Griffith Review* commissioned a dedicated quarterly issue, *First Things First,* and in its pages both Aboriginal and Torres Strait Islander thought leaders and their allies provided the backdrop to why the Statement was required (Schultz & Phillips, 2018). It detailed complex and divergent ideas on sovereignty, community-led approaches, the power of representation both politically and through modelling. In doing so it demonstrated that the Uluru Statement would not be singularly categorised as an unsupported solution, but that it forms an Indigenous-led ongoing conversation about self-determination. *First Things First* continued the conversation that the government refused, and by extending the discussion around our future possibilities it provided the Australian Government and Prime Minister, with some much needed edumacation.

Beyond these conversations, the challenges and aspirations remain in the work of Aboriginal academics like Bruce Pascoe, who has challenged previous non-Indigenous ideas of pre-invasion engagement with the land, in his book *Dark Emu* (Pascoe, 2016). It exists in the challenges that Aboriginal academic Maggie Walter makes in asking us to take on statistics to fight the government and academic apologists at their own game (Walter, 2018). These challenges are formed, and live, in the work of Aileen Moreton-Robinson, Martin Nakata, Karen Martin, Kaye Price, Marcia Langton and Bronwyn Carlson, as they prompt difficult conversations around identity, race and our right to succeed.

OUR ACHIEVEMENTS

An editor and contributor to the *First Things First* publication, Sandra Phillips, is an academic who is both an exemplary scholar and teacher. She is embedded in the aspirational story industry in which we must participate, bringing that knowledge to the academy. Sandra represents the reality for many of us as mid-career scholars, punching above the expectations set for our non-Indigenous counterparts. She has decades of experience as a publisher and editor at both Magabala Books and within UQ Press, is the Chair for the Centre for Indigenous Story, sits on multiple national arts boards and is engaged in edifying processes that promote agency for emerging writers and artists. And she accomplished this as a single mother raising three children (Phillips, 2018). Phillips' leadership and the change she effects challenges the criticism that those who succeed in our culture frequently hear both from outside of, and occasionally as Lerma contends, from other subjects of the colonial project (Lerma, 2014, pp. 124-126). She is connected and invested in both community and the academy. Her work informs embedding, her contribution provides modelling for students yet to come, and she is engaged in laying the groundwork for our publishing industry for years to come.

Leesa Watego is an Aboriginal business innovator and entrepreneur. She provides multiple opportunities for others in her work at Iscariot Media, in her commissioning of wearable thought pieces with Dark and Disturbing, as the founder of the Indigenous Business Program's nationwide networking initiative, Black Coffee, as the creator of Blacklines Publications producing Indigenous education resources, and as author of *The Critical Classroom* (Wilson, Carlson, & Sciascia, 2017). Her work is business focused, community targeted, and encompasses cultural, political, educational, economic and aspirational engagements, as she constantly provides training to other businesses and start-ups (SLQ Marketing and Communications, 2017). It is worth noting that she is a graduate of the Murra program.

Leesa Watego's work deeply informs education from her publications to the blogsites that she writes and promotes (Whatman, 2017), and while she is an occasional lecturer into higher education, her work defies the kind of structures that the academy understands, and she continues to educate people regardless of her place in the system. While we can celebrate her achievements, how do we, as a higher education community of practice, ensure that visionaries like Leesa become a mainstay of the learning process? We have accomplished this with academic Elders programs, where some universities have understood the enormous value that knowledge and leadership provided by Elders within our Communities brings. Now we must consider how we accomplish this for our business and industry leaders who will provide other, complementary paths to knowledge acquisition.

OUR FUTURE(S), OUR AGENCY

In the event of the Natives making the smallest show of resistance – or refusing to surrender when called upon so to do – the officers Commanding the Military Parties have been authorized to fire on them to compel them to surrender; hanging up on Trees the Bodies of such Natives as may be killed on such occasions, in order to strike the greater terror into the Survivors.

Governor Lachlan Macquarie orders to soldiers, 1818
(Macquarie, 1816–1818)

While this chapter tracks the barriers to change and agency, I contend that education and edumacation delivers strategies for us to challenge any past or present rejection of our cultures and our communities. That, of course, is just the beginning point. Our future is there in the *Uluru Statement from the Heart*, it is in the strategy work our leaders are accomplishing, it is present in the work of innovative entrepreneurs within our community. Our future is also in the collaboration with our fellow Australians, that will see us participating at every level and in every way that we can. And we *will* excel.

The caution is that in doing so, we must not forget our unique contribution, our differences and our values. We remain future focused, as only we can be, over our own destinies. And we replace reductive agreement that historically "we" were inventors through widely known – and often pan-Indigenised – tools like the yidaki (didgeridoo), the kylie (boomerang), our Songlines and our understanding of the galaxies; instead we scaffold this with our Knowledges of how we participated in ways that were never acknowledged. From Bruce Pascoe's (2016) detailing of complex systems of harvest, communication and understanding of the world to remembering the powerful Aboriginal and Torres Strait Islander People who came before us. Inventor David Unaipon, known to many Australians as the man on the 50-dollar note, whose work across his lifetime included the unacknowledged invention of radically innovative sheep shears. The work we do today in the academy is to support and promote these accomplishments, and to require acknowledgement, to hold the current and past systems accountable, and to ensure that the rest of Australia acknowledges the contribution of Unaipon and every other First Nations' inventor, scientist, thinker, doctor, teacher, artist, visionary and cultural leader.

Central to this chapter is the idea that our communities need to be at the forefront of leading change. In doing so, I have intentionally avoided placing specific responsibilities on communities, or suggested ways that they should be engaged. An ongoing conversation is just that, a moment in time when imperatives change and adjust. Mapping what these will be is less important in the landscape of our attainment in agency than ensuring we are not excluded from the pathways that will deliver what we need. This is the space of aspiration and possibility, and it is endless, but it is essential that it not be directed by an imperative for others to understand what we do and why we do it, to support us to determine, build and work with the tools that will support our agency.

ACKNOWLEDGEMENTS

This chapter is made possible by generations of First Nations' Peoples sharing their ideas about education, agency and the hope of a future where we control our own destinies. I acknowledge I come from the Wiradjuri; I am writing from Meanjin, on land of the Turrbal and Jagera Peoples.

NOTE

[1] First Nations, Aboriginal, Torres Strait Islander, Indigenous, Native, and Communities, where used in relation to First Nations' collective groups, are capitalised to demonstrate the short form for a proper noun as in Aboriginal Australian, First Nations of Canada or Wiradjuri Nation forming part of a specifically named collective (O'Sullivan, 2016, p. 71).

REFERENCES

Asmar, C., & Page, S. (2009). Sources of satisfaction and stress among Indigenous academic teachers: Findings from a national Australian study. *Asia Pacific Journal of Education, 29*(3), 387-401.
Australian Human Rights Commission. (2008). *Close the gap: Indigenous Health Equality Summit Statement of Intent*. Canberra, Australia: Author.
Australian Human Rights Commission. (2018). *Close the gap – 10-year review*. Canberra, Australia: Author.
Behrendt, L., Larkin, S., Griew, R., & Kelly, P. (2012). *Review of higher education access and outcomes for Aboriginal and Torres Strait Islander people* (Final Report). Canberra, Australia: Department of Industry, Innovation, Science, Research and Tertiary Education.
Bodkin-Andrews, G., & Carlson, B. (2016). The legacy of racism and Indigenous Australian identity within education. *Race, Ethnicity and Education, (19)*4, 784-807.
Bodkin-Andrews, G., & Craven, R. (2013). Negotiating racism: The voices of Aboriginal Australian post-graduate students. In R. Craven & J. Mooney (Eds.), *Diversity in higher education: Seeding success in Indigenous Australian higher education* (Vol 14, pp. 157-185). Bingley, England: Emerald Group.
Bond, C. (2014, November 14). When the object teaches: Indigenous academics in Australian universities. *Right Now*. Retrieved from http://rightnow.org.au/opinion-3/when-the-object-teaches-indigenous-academics-in-australian-universities/
Bourdieu, P. (2013). *Distinction: A social critique of the judgement of taste*. London, England: Routledge.

Evans, M. M., & Williamson, I. O. (2017). Understanding the central tension of Indigenous entrepreneurship: Purpose, profit and leadership. *Academy of Management Proceedings, 2017*(1), 14904.

Fforde, C., Bamblett, L., Lovett, R., Gorringe, S., & Fogarty, B. (2013). Discourse, deficit and identity: Aboriginality, the race paradigm and the language of representation in contemporary Australia. *Media International Australia, 149*(1), 162-173.

Fredericks, B. (2009). The epistemology that maintains White race privilege, power and control of Indigenous Studies and Indigenous Peoples' participation in universities. *ACRAWSA E-journal,* 5(1), 1-12.

Fredericks, B. (2013). We don't leave our identities at the city limits: Aboriginal and Torres Strait Islander people living in urban localities. *Australian Aboriginal Studies, 1*, 4-16.

Heckenberg, S., Gunstone, A., Anderson, S., & Hughes, K. (2016, April). *Nothing about us without us: Protecting Indigenous knowledges through oral histories and culturally safe research practices.* Poster presented at the Post Graduate Student Experience National Symposium, Gold Coast, Australia.

Janke, T., & Sentina, M. (2018). *Indigenous Knowledge: Issues for protection and management.* Retrieved from https://www.ipaustralia.gov.au/about-us/news-and-community/news/indigenous-knowledge-issues-protection-and-management

Janke, T. (2009). *Beyond guarding ground: A vision for a National Indigenous Cultural Authority.* Rosebery, Australia: Terri Janke and Co.

Jones, J. (2012). Dancing with the Prime Minister. *Journal of the European Association for Studies on Australia, 3*(1), 101-113.

Kovach, M. (2009). *Indigenous methodologies: Characteristics, conversations, and contexts.* Toronto, Canada: University of Toronto Press.

Langton, M. (2018). For her, we must: No excuses, time to act. *Griffith REVIEW, 60,* 328.

Lerma, M. (2014). *Indigenous sovereignty in the 21st century: Knowledge for the Indigenous spring.* Gainesville, FL: Florida Academic Press.

Lorde, A. (2018). *The master's tools will never dismantle the master's house.* London, England: Penguin Classics.

Macquarie, L. *Diary 10 April 1816 – 1 July 1818.* Original held in the Mitchell Library, Sydney, Australia.

Martin, K. (2003). Ways of knowing, being and doing: A theoretical framework and methods for Indigenous and Indigenist re-search. *Journal of Australian Studies, 76,* 203-214.

McKinnon, C. (2014). From scar trees to a 'bouquet of words': Aboriginal text is everywhere. In C. McKinnon, T. Neale, & E. Vincent (Eds.), *History, power, text* (pp. 371-383). Sydney, Australia: UTS ePRESS.

Moore, R. (2012). Whitewashing the gap. *International Journal of Critical Indigenous Studies, 5*(2), 2-12.

Moreton-Robinson, A. (2011). The white man's burden: Patriarchal white epistemic violence and Aboriginal women's Knowledges within the academy. *Australian Feminist Studies, 26*(70), 413-431.

Nakata, M. (1998). Anthropological texts and Indigenous standpoints. *Australian Aboriginal Studies, 2,* 3-12.

Nakata, M. (2010). The cultural interface of Islander and scientific knowledge. *The Australian Journal of Indigenous Education, 39*(S1), 53-57.

O'Sullivan, S. (2015). Stranger in a strange land: Aspiration, uniform and the fine edges of identity. In D. Hodge (Ed.), *Colouring the rainbow: Blak queer and trans perspectives: Life stories and essays by First Nations' Peoples of Australia.* Mile End, Australia: Wakefield Press.

O'Sullivan, S. (2016). Recasting identities: Intercultural understandings of First Peoples in the national museum space. In P. Burnard, E. Mackinlay, & K. Powell (Eds.), *The Routledge international handbook of intercultural arts research* (pp. 61-71), London, England, Routledge.

Pascoe, B. (2016). *Dark emu black seeds: Agriculture or accident?* Broome, Australia: Magabala Books.
Pascoe, B. (2018). The imperial mind: How Europeans stole the world. *Griffith REVIEW, 60*, 234.
Phillips, S. (2018). A rightful path: Educating for change and achievement. *Griffith REVIEW, 60*, 117.
Price, K. (2015). *Aboriginal and Torres Strait Islander education: An introduction for the teaching profession*. Cambridge, England: Cambridge University Press.
Referendum Council. (2017). *Uluru Statement from the Heart.* Retrieved from https://www.referendumcouncil.org.au/sites/default/files/2017-05/Uluru_Statement_From_The_Heart_0.PDF
Schultz, J., & Phillips, S. (Eds.). (2018). *Griffith Review 60: First things first*. Melbourne, Australia: Text Publishing.
SLQ Marketing and Communications. (2017, March 2). *Meet Leesa Watego: Entrepreneur, creative, academic, and teacher* (Web log post). Retrieved from http://blogs.slq.qld.gov.au/business-studio/business-studio-focus-on-indigenous-arts/
Universities Australia. (2017). *Indigenous Strategy 2017–2020*. Canberra, Australia: Author.
Walter, M. (2018). The voice of Indigenous data: Beyond the markers of disadvantage. *Griffith REVIEW, 60*, 256.
Whatman, S. (2017). Promoting wellbeing with educationally disadvantaged children through community partnerships. In S. Garvis & D. Pendergast (Eds.), *Health and wellbeing in childhood* (pp. 253-268). Cambridge, England: Cambridge University Press.
Wilson, A., Carlson, B. L., & Sciascia, A. (2017). Reterritorialising social media: Indigenous people rise up. *Australasian Journal of Information Systems, 21*. Retrieved from http://dx.doi.org/10.3127/ajis.v21i0.1591
Yu, A. C. L. (2004). Reduplication in English Homeric infixation. In K. Moulton & M. Wolf (Eds.), *Proceedings of the 34th North East Linguistics Society* (Vol. 34, pp. 619-633). Amherst, MA: GLSA.

Sandy O'Sullivan (Wiradjuri) PhD (ORCID: https://orcid.org/0000-0003-2952-4732)
School of Communication and Creative Industries
University of the Sunshine Coast, Australia

ROSEMARY LEONARD AND MARGOT CAIRNES

9. CHANGING WORK REALITIES

Creating Socially and Environmentally Responsible Workplaces

Given the task of discussing future workplace practices to further the common good, we were at first daunted by the current trends driven by powerful economic forces, which we describe in the first section. Like most leftist social scientists, we turned to look at government and intergovernmental bodies to moderate corporate behaviour, thus in the second section we reflect on how the workplace would be managed for the environment and social capital if governments had the ability, and the will, to manage workplace behaviour within a capitalist system. Although these are nice thoughts, we argue that this is an unlikely scenario. In the third section we examine potential alternatives to the current dominant paradigm: that is, workplaces that operate outside capitalism. Our argument is that these alternatives do already exist, albeit in comparatively small numbers, and provide flexibility for companies to contribute to the common good because they are freed from the tyranny of the need for growth. The development of alternatives needs to blossom now to demonstrate the potential for change and give hope for meaningful work to those rejected by, or rejecting, the capitalist workplace. Finally, we find hope in the trends to re-localisation whereby local communities focus on the resources at hand to generate work and fill their needs. We are not arguing for a return to parochialism however, because local places are now connected to the world through the internet and are thus part of global networks. However, local dependence requires fostering and growing our human and social capital and cherishing our environmental capital, thus building workplaces for the common good for people, and the natural world.

CURRENT TRENDS FOR THE WORKPLACE

There is no doubt the world of work is changing. With digital disruption, for example, we can expect 40–70% of current jobs to disappear within the next 10 years. Social interaction has been altered by numerous developments such as electronic meetings, distance working, global virtual teams and social media. Further, artificial intelligence means that machines are learning so fast that the nature of work is hard to predict.

All these changes are occurring within a capitalist system which depends on economic expansion because at its most basic, capitalism is about obtaining credit in the present with the expectation of future profit. The issue is not so much that corporations need to make a profit, but that they need to make increasingly large profits. It is a system which allows the rich to become richer. On the one hand, supporters of the system say that this benefits everyone due to the "trickle-down effect". They also point to the substantial donations made by corporations under the banner of corporate social responsibility (CSR). On the other hand, critics point out that the poor are becoming poorer and that CSR contributions are usually tax deductible and a very small proportion of corporations' wealth.

Most developed countries are forms of neo-capitalism. This means that governments maintain some control over corporations and provide a safety net for those missing out on the benefits of capitalism. Further, most governments have adopted neo-liberalism, which is broadly defined as the extension of competitive markets into all areas of life, including the economy, politics and society (Springer, Birch, & MacLeavy, 2016). The move to neo-liberalism seems to have occurred with the propaganda victory for capitalism which claimed that "there is no alternative" (TINA). Neo-liberalism encourages smaller government and pressure on the watch-dogs of corporations which enables corporations to externalise more and more of the negative consequences of their activities for people and the environment because there are insufficient government resources to monitor and prosecute corporate misdeeds. Small government also means that the safety nets that governments provide are becoming more constrained and dependent on workforce participation at the same time as work is disappearing.

So what does all this mean for the workplace? Over 20 years ago Charles Handy (1994) wrote about the three-tiered nature of work: the rich corporate employees; the portfolio workers (managing a variety of contracts); and the itinerant or piece workers who, in many countries, lack any backup or safety net. With current trends in the development and adoption of automation and artificial intelligence there will be fewer corporate employees, more (but poorer) portfolio workers, and even more piece workers who are virtually the hidden unemployed. This seems to have been a fairly accurate prediction and more so for countries where there is little in the way of corporate regulation or a safety net. Another trend is that project and piece workers are increasingly expected to become small businesses which are contracted by large firms rather than employed by them. This means that workers take all the responsibility for their work conditions such as sick leave, superannuation, training, equipment and insurance.

Although governments are important we would argue that the not-for-profit (NFP) sector can also have a role in shaping the future of the workplace. The NFP sector consists of all those formal and informal groups and organisations ranging from large charities, which increasingly seem to look and behave like corporations, down to informal loosely connected groups of people that come together for some purpose. NFP organisations are seen positively by corporations and right-wing governments when they manage the externalities of capitalist society (e.g. the negative consequences for people and the environment) and provide services at a cheaper rate than government which reduces the pressure for higher taxes. Such organisations usually rely on government funding and therefore are easily controlled. However, NFP organisations can also be seen negatively and often pressured by corporates and right-wing governments when they focus on advocacy and lobbying for people or the environment. These include groups which keep watch on the corporate and government sectors looking out for environmental breaches, corruption or inhumane practices, and also unions, guilds and other professional bodies which look out for workers' conditions and professional standing. Such organisations play an important role in identifying corporate misdeeds even if they are increasingly less able to stop them. However, the most valuable aspect of the NFP sector for the future workplace might stem from its history of innovation. Indeed, many of the government systems we take for granted today, such as health, education and welfare, were first enacted in the NFP sector. Thus, the sector also includes a motley collection of community development groups who find new ways to get things done, some of which do not involve obtaining credit and repaying the debt with interest, i.e. they are outside capitalism.

How these current trends play out for future workplaces depends on the responses and interactions of the three sectors: corporate, government and NFP. In this chapter we focus on two dynamics which could make workplaces facilitators of the common good:

1. Enforcing the triple bottom line – governments and international governing bodies maintain or increase their power over corporations including the power to enforce TBL accounting and performance indicators from companies.
2. The death of TINA – multiple alternative ways of getting things done emerge and are recognised as positive ways of being part of our social fabric.

DYNAMIC 1: ENFORCING TRIPLE BOTTOM LINE ACCOUNTING AND PERFORMANCE INDICATORS

Traditional left-wing political thought focuses on the role of democratically elected governments to act for the common good of the population. Left-wing thinkers and activists call on governments and international governing bodies to increase their power over corporations, including the power to force companies to behave in socially and environmentally responsible ways. This is often presented in terms of TBL accounting (accounting for social and environmental impacts as well as financial outcomes) and performance indicators. Of the three dimensions, monitoring is often best-established for finance. For the second dimension, environmental responsibility, we believe that the path to monitoring and control is reasonably clear. Initiatives that could be adopted include: a carbon price and carbon market; assessment of all installations for whole of life cycle carbon emissions (Tam et al., 2017); company funding for the rehabilitation to any land damaged while they were operating; better monitoring of existing regulations; and prosecution of breaches. For the environmental dimension, governments know what needs to be done, they just need the political will to do it. However, the situation for the social dimension of the TBL is not so clear, so in the next section we explore how workplaces can become drivers of social capital.

Creating Social Capital in the Workplace

Although social capital has a variety of definitions, at its core it is a resource that accrues from relationships – networks and trust (Bourdieu, 1986; Leonard & Onyx,

2004; Putnam, 2000) and there is growing recognition of its importance in the workplace. As early as 2000, Onyx and Bullen (2000) found a workplace subscale in their social capital scale and recently Eguchi et al. (2017) identified six elements of workplace social capital which covered: keeping each other informed; a "we are together" attitude; feeling understood and accepted; helping each other; trust; and laughter and smiles. Evidence is accruing that social capital has multiple benefits in the workplace for the mental health of workers, their life satisfaction (Requena, 2003) and the effectiveness of their work (Gant, Ichniowski, & Shaw, 2002; Oh, Chung, & Labianca, 2004). However, while social capital cannot be created from above, the conditions for fostering social capital can be (Leonard & Onyx, 2004). Basically, those conditions are opportunities for ongoing substantial connections in a non-competitive and ideally cooperative way.

Workplaces have traditionally been fruitful avenues for developing relationships. At the most basic level they provided opportunities for informal interaction over an extended period so friendships could grow if people were so inclined. Opportunities for informal interaction were aided by regular timetables, common areas for meals, and staff social functions. In recent years increased flexibility has been favoured by workers and management. Flexible working hours, working from home, eating at the desk, and workplaces without separate kitchens so there can be no talk while getting a glass of water or cup of coffee, all reduce opportunities for social networks to form. Web-based work and surveillance technology allow even more flexibility and less time in a common place.

The development of social capital requires more than a passing acquaintance however: it needs trusting relationships. Trust can be of a specific nature whereby we only need to trust a person to perform their role competently and honestly. More complete trust requires a more thorough knowledge of the person. Informal office chatter helps people obtain a richer knowledge of each other, perhaps identifying common interests or experiences, knowing who you can talk to about a difficulty, getting together outside the workplace, or collaborating for charity fundraising. These commonplace interactions are all part of developing multiplex relationships and denser networks.

Workplaces can easily design in opportunities for informal connections but people will not take up those opportunities unless they think it is worth the trouble. Management could actively facilitate these connections and relationships. For example, they could highlight goals for which people need to work together and allot rewards and recognition to groups rather than individuals. Teams which are small enough for people to actually get to know each other are valuable but there is a risk that this can set up divisions within the workplace (Andersen et al., 2015) especially if they are in competition with each other. Therefore, it is desirable to have teams working on different tasks so they are not in competition and multiple team involvements so people get to know others beyond their own team.

Effective mobilisation of social capital requires networks of trust at all levels. Indeed, Oksanen et al. (2010) demonstrated that for workplace social capital, the vertical component (i.e. respectful and trusting relationships across power differentials at work) and the horizontal component (trust and reciprocity between employees at the same hierarchical level) were both important for workers' mental health. Further, Helliwell and Huang (2010) found that a one-third standard deviation increase in trust in management is equivalent to an income increase of more than one third for improving life satisfaction, and Sapp et al. (2010) showed that trust in managers mitigated the negative effects of high workplace stress. However hierarchical structures make it difficult for those higher up to see what happens at the coalface. Middle managers can behave like clouds – all bright and shiny for those above, black and ominous to those below, and opaque so the one cannot see the other. When managers lack personal relationships with workers, they are more likely to have negative beliefs about their performance and the need to enforce productivity (Kipnis, 2008). These structures can allow bullying, harassment and favouritism to thrive and managers to "big note" themselves, hide poor performance and then move on before they are found out (Gillespie, 2017). Dishonesty and lack of disclosure by those with power, management resistance to staff associations, and attempts to drive down working conditions give the general message that people are of no value at all and destroy trust. In contrast workplace leaders could demonstrate high levels of integrity and respect for people at all levels. Organisational practices such as setting up multiple communication channels between all the levels of management and widely distributing decisions make it difficult for people to "play games" at the expense of others.

Organisations who value social capital also recognise the importance of building social capital beyond the workplace. The InLoop case study, particularly their regional site, is a good example of complementing the bonding social capital within the company with bridging social capital with the community. Because both bridging and bonding social capital support the development and appreciation of trusting relationships it is likely that they are complementary unless there is some concern about connecting with the wider community, e.g. minority groups experiencing discrimination (Leonard & Bellamy, 2010).

Social capital is often strongest when there are multiplex relationships, that is, when people are connected in a number of ways, e.g. workmates can be friends and community group members (Leonard & Onyx, 2004); however, such multiplexing is difficult when people need to create workplace personas which are at odds with the presentation of self in other contexts (Goffman, 1959). A further barrier to social capital is a more generalised de-personalisation of human relations that is triggered by certain workplaces. For example, Colbert, Yee, and George (2016) identified the lack of empathy growing in the digital workforce as people are dealt with simply as data.

In our first case study we discuss the company InLoop. InLoop value bridging social capital and take corporate social responsibility seriously, not just to enhance their image and increase sales but also to strengthen the wider community. Organisations such as InLoop which value bridging social capital need a high degree of transparency because social capital requires trust in their operations. They need to behave as responsible citizens – not just complying with the letter of the law while searching for every legal loophole to externalise any damage to people and the environment. They would have organisational goals which relate to the public good (not just profit) and enact them. The case study is based on an interview with Steve Austen by the second author that was first published online in *CXFocus Magazine* (Cairnes 2017b; used with permission).

Case Study 1: InLoop

InLoop is one of Australia's fastest growing start-ups and winner of several business growth awards including the BRW Fast100 in 2013 and 2014. It is a financial technology company, founded in 2005 by three young men with complementary skills. Steve was the entrepreneur, Chris was the IT guru and Geoff was a young retired executive from Macquarie Bank, bringing both funds and financial savvy. The company provides payment platform services to a range of industries including government, education, health and the private sector. The company services 650,000 active customers and manages over 30 million transactions annually. Steve is the prototypal IT success story. Steve's office in Kingscliff is 500 metres from his home and directly overlooks Kingscliff Beach on the NSW Far North Coast. Most days, Steve walks to work dressed in board shorts and t-shirt: "I start at 7.30am and generally work until 5.30 or 6pm to have dinner with the family" (Cairnes, 2017b, n.p.).

InLoop recognises the value of employee involvement and commitment. They have two offices in Sydney where the atmosphere is electric with creativity and excitement: "the really good guys come for the stimulation of being with each other ... We only hire people who are people oriented – it is extremely hard to find people with technology skills who are people-centred in their thinking. They cost twice as much but they are 20 times more effective" (n.p.).

It is in their regional Kingscliff office that the effects of social capital are most obvious. The Kingscliff office has been relocated three times to accommodate the growing number of employees. And the company in turn shows a commitment to staff: "Our office works really hard but we have a laugh all the time. The staff appreciates getting out at lunchtime and having their feet in the sand. It is a great way to recharge" (n.p.). Nobody works on weekends unless it is absolutely essential. The Kingscliff office is open plan, very colourful and supplies health snacks in its vibrant kitchen, a popular meeting place. InLoop has regular social activities in all its offices and with all its offices together. In Kingscliff these often include whole families. Social capital building also extends beyond the office. InLoop believe that a company in a regional area becomes part of the community and so people feel much more engaged with the organisation. Not only is InLoop a major employer in Kingscliff but they are also a significant employer of local consultants and contractors. Further, the workers are encouraged to be active in the local community, schools and charities, and recently InLoop's whole office volunteered at an event for families with autism. As Steve puts it: "we are building a community not just a business" (n.p.).

Will Triple Bottom Line Accountability be Enforced?

In this section of the chapter we have looked at government enforcement of TBL accounting as a path to ensuring more positive future workplaces. In particular we have examined the social dimension which is the least developed in terms of monitoring and accounting. We end the section by asking how probable it is that governments and inter-governmental agencies will take the necessary action to enforce TBL accounting. There are trends that would support the re-emergence of government power over corporations and their ability to enforce TBL accounting. First, the shift might be prompted by increasing volatility in the market. Companies need to play the capitalist game and that requires them to play by the rules at least to a certain extent – otherwise the market loses confidence which then triggers a downturn. Governments have an important role in creating the rules. Second, a shift to a stronger government position might occur because of their buying power. Increasingly, governments may become key customers because they are large enough to buy the huge products that companies need to sell to expand. Third, companies might recognise that because it is in their interests to have an educated, healthy, cooperative workforce, they might need to provide governments with some support. A fourth imperative to the enforcement of TBL accounting might come through the legal system. For example, loss of employment and workplace bullying are major causes of suicide and mental breakdown. Successful legal action by people and families who have experienced these losses might encourage monitoring of the social dimension of the TBL.

On the other hand, there are a number of forces which make increases in government control seem unlikely. Governments are too committed to the neo-liberal model or too scared of the consequences of reining in corporations. In Australia, the mantra of "Jobs and Growth", implying growth is necessary to maintain employment levels, applies to both the major political parties although Labor are more concerned with jobs and Liberals with growth. Because capitalism depends on expansion, corporations have only two modes – expansion or collapse – and governments may be legitimately scared to rein in corporates if that causes them to move their business to another state or nation or triggers a major recession bringing widespread misery. Therefore, we are not optimistic. Forcing social and environmental responsibility on corporations would require high levels of commitment, monitoring and inter-governmental cooperation which we just cannot see happening in the near future.

DYNAMIC 2: THE DEATH OF TINA

If we cannot rely on governments to make companies behave in more socially and environmentally responsible ways then we need to create alternative workplaces – places where people want to work and can feel proud of their contribution to community and the environment. The idea that "there is no alternative" is both powerful and oppressive and can best be overcome by demonstrating that there are alternatives. In this section we examine two trends: Paul Mason's conception of post-capitalism and re-localisation.

Post-capitalism

Already there are signs of alternatives to capitalism emerging. People are finding other ways to get things done which do not involve obtaining credit and repaying the debt with interest, i.e. they are outside capitalism. Paul Mason shows that new forms of ownership, new forms of lending, new legal contracts and a whole new business subculture have emerged over the past 10 years. The new systems, which the media has dubbed the "sharing economy", involve collaborative production, using network technology to produce goods and services that only work when they are free or shared. Mason argues that these innovations mark the route beyond the market system and that networks can be the basis of a non-market system that replicates itself (Mason, 2015). He uses the example of when Greece defaulted on its fiscal agreement with the European Union. Austerity measures were causing distress across the country but people responded with a wide range of projects to fulfil community needs including food co-ops, alternative producers, parallel currencies and local exchange systems. There were at least 70 substantive projects and hundreds of smaller initiatives ranging from squats to carpools to free kindergartens. The focus for the Greek communities was working out local systems for which the concept of default was not relevant. They wanted to escape the system which was punishing their nation with austerity measures (Mason, 2018).

Mason (2018) asserts that the state will need to create the framework for new post-capitalist systems of the sharing economy. Although a state supported framework would be useful we think it is unlikely to eventuate in the near future. What we see at present is that governments are unsupportive of alternative systems and try to herd everyone into the capitalist corporate model. So we do not think we can wait for governments to provide useful structures for post-capitalist activities; in the meantime, to provide positive vibrant workplaces for the future we will need to get on and create them anyway. In the case study below, we show how Flow Hive has managed to thrive without being driven by the growth required to repay debt. It has also identified the value of good social and environmental practices for financial success. Thus, for Flow Hive the three dimensions of the triple bottom line are intermeshed. The case study is based on an interview with Cedar Anderson by the second author that was first published online in *CXFocus Magazine* (Cairnes 2017a; used with permission).

Case Study 2: Flow Hive

Cedar Anderson, co-inventor and co-leader of Flow Hive is a great example of a new breed of entrepreneurs who are choosing independence and the right to contribute to the common good. Cedar notes: "The new way of thinking is you do business for positive purpose. In the old way you make as much money as you can and don't care how. You then give a bit away at the end so you don't feel so bad.

The new way is, from square one, what you are doing makes a positive impact" (Cairnes, 2017a, n.p.).

Cedar and his father Stuart are beekeepers from Byron Bay, in northern NSW, who developed a hive that allowed honey to be harvested without opening the hive. This innovative method reduces extraction times and stress on the bees because it eliminates the need to smoke them, dismantle the hive and remove the honeycomb. Cedar and his dad Stuart decided to crowdfund their innovation through Indiegogo. When they launched on 23 February 2015, Indiegogo crashed several times from people refreshing. They reached their target of $70,000 in seven minutes. They went on to raise $US12.2 million in six weeks. Two years on, Flow Hives are being distributed globally by Amazon, eBay and other online traders.

Because of their internet success Cedar and his dad did not need to borrow from a bank, meaning they had no loans to repay, no interest to service and no external shareholders to satisfy. This way they were free from the force of expansion or failure (boom or bust) and thus free to stay true to their heartfelt values. The ease of harvesting honey using Flow Hive has led to a rapid rise in the number of beekeepers around the globe. Flow Hive has encouraged this trend by developing a support team to answer questions by email, phone and live on Facebook.

They use their positive media presence to promote causes close to the hearts of their followers, for example, helping victims of the Vanuatu cyclone and the Nepal earthquake: "All along the way we are trying to think creatively – how can we have a bigger impact beyond just beekeeping? There is so much more we can easily do" (n.p.). For Cedar, doing good, living your values, and being successful in business are synonymous: "Our aim is that that even when we are dead and gone you can't shake off the positive world stuff or the brand would become worthless. A lot of companies are cottoning on that they need some aspect of social enterprise or positive environmental impact or they just won't be in the game" (n.p.).

Re-localisation

Re-localisation is a movement which recognises our wastefulness and our vulnerability arising from a dependence on complex global networks for even our most basic needs. For example, the provision of food, water and waste systems are all highly dependent on oil-based transport (Friedemann, 2016). Re-localisation is a response to that threat which focuses our attention on local resources.

The Transition network is a strong advocate for re-localisation. Transition has been a rapidly growing worldwide network with approximately 1,000 initiatives which learn and experiment with new ways of operating. Their principles include respect for the limits of the planet, social justice, subsidiarity (self-organisation and decision making at the appropriate level), positive visioning, sharing ideas and power, and working collaboratively to unleash collective genius. They support "REconomy" to ensure that local economies support the changes we want to see in the world. Initiatives include community-owned energy companies, food businesses, local currencies, helping existing businesses to change their models, and improving access to the right types of skills and investment (Transition

Network, 2018). Transition is a good example of the use of the internet. Local groups can now access high levels of international expertise. So the call for re-localisation is not a regression to basic subsistence levels. The technology that created computers, the internet and wifi is now widely known and provides us with the latest knowledge which allows us to "think globally while acting locally".

Effective re-localisation requires the identification and mobilisation of skills and resources of a local area and often that means disregarding sector boundaries and asking what people from each sector are able and willing to bring to the table to support a local initiative. Case Study 3 presents the second author's reflections on her experience in a community organisation promoting local development by recognising the human capital of its residents.

Case Study 3: Sourdough, Byron Bay

Sourdough is a community networking organisation which is promoting local business and community development and which provides an interesting example of mobilising local knowledge and skills to support the local economy and community. The last detailed study of the region (Leo, 1999) found that it had the second highest level of unemployment for any region in Australia. Around 95% of businesses employed fewer than 10 people. Unemployment amongst young people was high (25–30%) and the rate amongst the Aboriginal population was extremely high. There was also substantial underemployment and hidden unemployment problems. Personal experience in the community suggests that, in recent years, this profile has been exacerbated by the rising cost of living, especially housing.

Started by a group of highly successful business leaders who had made a sea/tree change to the Northern Rivers, Sourdough draws on the years of skill and experience of those who have moved away from the "hurly burly" of big capitalism. Sourdough has amassed a group of over 180 "Hidden Assets", which is their name for the refugees from the corporate world of intense competition who have chosen to live a more community-based life. Sourdough seeks to use the skills of its Hidden Assets to mentor and support business owners and up-and-coming entrepreneurs to build their businesses and thus create employment opportunities in the area. Apart from being very helpful to local business, Sourdough creates a forum for skilled and experienced professionals and business refugees to socialise and keep abreast of trends in both the region and the wider world.

It is a life-giving resource to both local business and the Hidden Assets. For example, recently Sourdough Women – a subgroup of Sourdough – took a group of around 30 women through U-lab, a course on business and social change. U-lab encourages all organisations and workers to think about engaging all stakeholders in problem solving, thus creating strong connections inside and outside the workplace. It is about creating a new source of social capital. Sourdough Women is currently working with a Google Project Manager to create an app based on key learning from the course to make available to women worldwide. Sourdough uses funding from foundations and partners with Byron Community College to support community projects including one to maximise Aboriginal health and wellbeing.

CONCLUSION

In this chapter we have argued that the capitalist requirement for ever-increasing profit is creating workplaces that are toxic for people, their communities and the environment. In order to create workplaces which promote the common good we need to create alternatives to the current neo-capitalist trends. Increasing numbers of people are being made redundant by automation or work intensity that is not sustainable or simply leave because they cannot accept the workplace culture or mindlessness of their work. Others will be under-employed or in insecure employment so they will not be worthy of credit. One positive outcome of people's marginalisation from capitalism is that they will have more time to put into the creation of alternative systems. It is likely to be the "Cultural Creatives" (Ray & Anderson, 2000) who will be least able to tolerate the corporate workplace but will be invaluable for generating alternatives. We have also argued that we cannot wait for governments to regulate the changes that are needed. With international collaboration, they could potentially enforce triple bottom line standards, ensuring good social and environmental outcomes, but they are too embedded in neo-liberalist thought and too concerned about economic downturn.

Hope arises from the many workplaces and communities that are already developing alternatives, such as businesses that are valuing their workers and building community, values-based businesses that reject the capitalist essential of growth based on credit, or NFP organisations that support re-localisation at the global or local scale. The examples we have provided are by no means the only ones. We see re-localisation as a particularly fruitful path to post-capitalism because it necessitates the valuing of local people, social networks and natural resources, hence it needs to maximise social capital and environmental sustainability. Coordination among the three sectors is desirable and we believe that governments, especially local governments, will start to support the changes that are happening anyway. However, it is people, their businesses, and their NFP organisations that are leading the way.

REFERENCES

Andersen, L. L., Poulsen, O. M., Sundstrup, E., Brandt, M., Jay, K., Clausen, T., Borg, V., Persson, R., & Jakobsen, M. D. (2015). Effect of physical exercise on workplace social capital: Cluster randomized controlled trial. *Scandinavian Journal of Public Health, 43*(8), 810-818.

Bourdieu, P. (1986). The forms of capital. In J. G. Richardson (Ed.), *Handbook of theory and research for the sociology of education* (pp. 241-258). Westport, CT: Greenwood.

Cairnes, M. (2017a). Flow Hive: Doing business for positive purpose. *CXFocus Magazine.* Retrieved from http://www.cxfocus.com.au/executive-profiles/flow-hive-doing-business-for-positive-purpose/

Cairnes, M. (2017b). Steve Austen from InLoop is living the start-up dream. *CXFocus Magazine.* Retrieved from http://www.cxfocus.com.au/executive-profiles/steve-austen-from-inloop-is-living-the-start-up-dream/

Colbert, A., Yee, N., & George, G. (2016). The digital workforce and the workplace of the future. *Academy of Management Journal, 59*(3), 731-739.

Eguchi, H., Tsutsumi, A., Inoue, A., & Odagiri, Y. (2017). Psychometric assessment of a scale to measure bonding workplace social capital. *PLoS One, 12*(6), e0179461.

Friedemann, A. J. (2016). *When trucks stop running: Energy and the future of transportation*. Cham, Switzerland: Springer.

Gant, J., Ichniowski, C., & Shaw, K. (2002). Social capital and organisational change in high-involvement and traditional work organizations. *Journal of Economics and Management, 11*(2), 289-328.

Gillespie, D. (2017). *Taming toxic people: The science of identifying and dealing with psychopaths at home and at work*. Sydney, Australia: Macmillan.

Goffman, E. (1959). *The presentation of self in everyday life*. New York, NY: Anchor Books.

Handy, C. (1994). *The empty raincoat: Making sense of the future*. London, England: Arrow Books.

Helliwell, J. F., & Huang, H. (2010). How's the job? Well-being and social capital in the workplace. *ILR Review, 63*(2), 205-227.

Kipnis, D. (2008). Does power corrupt? In J. M. Levine & R. L. Moreland (Eds.), *Small groups: Key readings* (pp. 177-186). New York, NY: Psychology Press.

Leo, A. (1999). *Case study on employment and unemployment in the Northern Rivers region of New South Wale* (Jobs for the regions: A report on the inquiry into regional employment and unemployment: Government response). Canberra, Australia: Commonwealth of Australia.

Leonard, R., & Bellamy, J. (2010). The relationship between bonding and bridging social capital in Christian denominations across Australia. *Nonprofit Management and Leadership, 20*(4), 445-460.

Leonard, R., & Onyx, J. (2004). *Social capital and community building: Spinning straw into gold*. London, England: Janus Publishing.

Mason, P. (2015, July 17). The end of capitalism has begun. *The Guardian*. Retrieved from https://www.theguardian.com/books/2015/jul/17/postcapitalism-end-of-capitalism-begun

Oh, H., Chung, M-H., & Labianca, G. (2004). Group social capital and group effectiveness: The role of informal socializing ties. *Academy of Management Journal, 47*(6), 860-875.

Oksanen, T., Kouvonen, A., Vahtera, J., Virtanen, M., & Kivimäki, M. (2010). Prospective study of workplace social capital and depression: Are vertical and horizontal components equally important? *Journal of Epidemiology and Community Health, 64*(8), 684-689.

Onyx, J., & Bullen, P. (2000). Measuring social capital in five communities. *The Journal of Applied Behavioral Science, 36*(1), 23-42.

Putnam, R. (2000). *Bowling alone: The collapse and revival of American community*. New York, NY: Simon & Schuster.

Ray, P. H., & Anderson, S. R. (2000). *The cultural creatives: How 50 million people are changing the world*. New York, NY: Harmony Books.

Requena, F. (2003). Social capital, satisfaction and quality of life in the workplace. *Social Indicators Research, 61*(3), 331-360.

Sapp, A. L., Kawachi, I., Sorensen, G., LaMontagne, A. D., & Subramanian, S. V. (2010). Does workplace social capital buffer the effects of job stress? A cross-sectional, multilevel analysis of cigarette smoking among U.S. manufacturing workers. *Journal of Occupational and Environmental Medicine, 52*(7), 740-750.

Springer, S., Birch, K., & MacLeavy, J. (Eds.). (2016). *The handbook of neoliberalism*. Abingdon, England: Routledge.

Tam, W. Y. V., Senaratne, S., Le, K. N., Shen, L. Y., Perica, P., & Illankoon, I. M. C. S. (2017). Life-cycle cost analysis of green building implementation using timber applications. *Journal of Cleaner Production, 147*, 458-469.

Transition Network. (2018). *Transition Network: About the movement*. Retrieved from https://transitionnetwork.org/about-the-movement/

Rosemary Leonard PhD (ORCID: https://orcid.org/0000-0002-7642-5529)
School of Social Sciences & Psychology, Western Sydney University, Australia

Margot Cairnes BEd (Hons.), MBA Founder: 12 Steps for Business, Australia

MEGAN CONWAY AND JOY HIGGS

10. TOWARDS FUTURE PRACTICE IN SOCIO-POLITICAL CONTEXTS

This chapter examines socio-political factors underpinning the future of practice and explores how to better position practice futures to respond to these contexts and factors. It proposes that the ability of young people (particularly) to position themselves and build possible futures for themselves remains contested and constrained by competing policy, cultural, social and economic interests. It examines the factors impacting on practice and explores possible supports to mitigate these factors for individuals, employers and educational bodies. We examine factors that might influence possible life and practice outcomes and pathways of individuals desiring positive work futures and examine these key questions:

- What are the key socio-political factors influencing future practice?
- What implications do these factors have for practice and work broadly?
- What recommendations might be considered to enable choice and agency for workers and practitioners amidst contested socio-political landscapes?

Across Western democracies like Canada and Australia there has been an increasing emphasis on what the future of work could look like, especially for younger people who are either midstream in education or practice, or are in the process of identifying an education or employment pathway. A scan of newspaper headlines illustrates a fractured discourse related to what might be looming on the work horizon. For example: What role will technology and artificial intelligence (AI) play in future workplaces (see Davis, 2018; Hancock; 2018; Howcroft & Rubery, 2018; Minku & Levesley, 2019)? What are the risks of outsourcing employment opportunities (Biddle, Sheppard & Gray, 2018)? What roles should universities and colleges play in preparing for tomorrow's workforce (Jackson et al., 2016; Kak, 2018)?

Questions without clear answers, and sometimes without attempts to answer, dominate this discourse around workplaces, people entering or re-entering the workforce, and the educational bodies preparing them to participate in the workplaces of the future. The shifting socio-political context in which work and education, and specifically practice, is situated is problematic. Practice and work are constantly being impacted by a range of global-economic and socio-political influences such as regional destabilisation due to war and economic recession (Humphreys, 2003; Mansfield & Pollins, 2003), mass migration of refugees (Chatty, 2010; Ferris & Kirişci, 2016), changing political alliances (Smith, 2011), climate change policies and challenges (D'Olympio, 2015), the crisis of contemporary democracy (Held & Hale 2017), changing global political powers

(Nye, 2004), the Fourth Industrial Revolution (Schwab, 2017), major technological advances particularly in the communications field, and the integral links between information technologies and global power (Rosenau & Singh, 2002).

> Simply put, people's lives are increasingly affected by international networks, operating via financial markets, transnational corporations, and the Internet, that impinge on traditional seats of authority and meaning, such as family, ... and nation. In response to the increased power of global networks, people – as individuals and in collectives struggle to assert control over their identity and defend what they see as essential to that identity. (Warschauer, 2000, p. 512)

Both socio-political economic contextual factors and technological changes hold the ability to mitigate and inhibit choice and agency for students and future workers. For example, while mass education policies, built on democratisation of education and the social justice agenda in many countries (Altbach 1992), have made higher education (and therefore graduates' range of educational and practice opportunities) accessible to a wider proportion of populations, changes in graduate employment prospects and changes in university funding have added enormously to graduate debt (Heller & Callender 2013), while limiting career options, making prospective students think twice about pursuing university education.

FACTORS INFLUENCING WORK AND PRACTICE

The challenge (while not necessarily the *problem* with practice future(s)) is their possibilities as well as their unpredictabilities; that is, that they are unknown. The following summarises some of the factors that are likely to influence the future of education and the workforce – and ultimately practice. First, factors that influence work, practice and (consequentially) education are not value neutral. They are mediated and informed by context as well as by the individuals who are informed by and are informing them. Practice norms and dominant discourses as to what is acceptable, influence work, practice and education and how these are perceived and influenced by each individual person and context (see Kalbfleisch 2003).

Second, influences on work and practice occur at different levels, often with differences in the directness of impact on people. The global shaping of a profession, for instance, is likely to have less immediate impact than whether your direct supervisor has a "personality conflict" with you. Often the more meta influences are political and economic while the more local influences are social.

Third, the modern spheres of work are influenced by a number of social, demographic and interpersonal factors. Looking through a social lens, how work and education are understood is strongly influenced by the construction of individual, family and community identities (Brown et al., 2009). These narratives, of what it is to work or to go to school, are shaped at an early age with expectations and a sense of *what should be* in terms of life outcomes. These are also strongly influenced by culture and what is expected by one's dominant cultural norms. The presence of role models, supports and behavioural norms related to work and educational identities are formed early in life and reinforced through decisions made in school and workplace settings.

Place

Place can strongly influence work and available opportunities. The reality of a place-based approach to considering work and practice is that a place-based analysis cannot be homogenous or uniform – there are thriving inner cities and faltering ones, prosperous rural communities and those that are struggling. According to a recent article by Alan Flippen (2014), data points such as education, median household income, unemployment rate, disability rate, life expectancy and obesity can be examined and spatialised to identify areas in the United States that are healthy, wealthy or struggling. As neighbourhoods and cities either build their labour market infrastructures or fail to make these investments, a corresponding pattern of labour market growth or decline is emerging in rural and urban areas across countries like the United States (Florida, 2018). The strong shift towards urbanisation and a retreat from rural centres has meant declines in available jobs and educational opportunities in rural communities. This is especially true in countries like Canada and Australia (Duffy-Jones & Argent, 2018; Malatest & Associates Ltd, 2002). This shift in labour market growth has meant a decline in practice opportunities in certain communities as capacity in these locales remains stretched and limited to address basic need. For example, how might a hospital that has identified a need to upgrade its nurse training and expertise in cardiac care, for instance, build this capacity in a remote and rural locale? Expertise often exists in urban areas that are a far distance away from smaller communities, and physical and cultural barriers often inhibit acquiring this knowledge/expertise. Further, governments often fail to consider how *place* impacts the delivery of education, work and practice opportunities, and as such there are frequently practical and cost limitations that present barriers to improving local knowledge and capacity.

Demographics: Gender, Age, Class, Culture

Demographic factors including gender, age, class, ethnic similarity, diversity and culture play a significant part in influencing the way people work and the way they are treated in relation to work including the nature of working relationships, tenure, success, discrimination, harassment, opportunities, progression and job satisfaction, team work and organisational effectiveness (see Epitropaki & Martin, 1999; Gonzalez & Denisi, 2009; Mayer, 2004; Riordan & Shore, 1999).

Actors and Stakeholders

Actors (i.e. people directly working in a workplace) and stakeholders (those people and groups within and outside of an organisation/work context who are influenced by or who influence the operations and actions of the workplace/organisation), all impact on the nature of practice/work and will continue to do so.

Global and Local Influences

Global and meta-level influences are those that impact at a national or culture-wide level, for instance. They particularly focus on broader economic, political, technological and labour market forces that are shaped or directed by government policy and knowledge/technology advances as well as globalisation. This level of influence also includes cultural and society-wide norms such as socially constructed and shared assumptions about employment and work, the relative importance of the public versus private good (Baston, 1994) and the impact of social capital considerations on government decisions (Boix & Posner, 1998). For instance, there is a considerable variation across professions, cultures, genders and organisations as to acceptable work-life balance and shared family versus employment workloads (see Gambles, Lewis, & Rapoport 2006). At organisational, industry and sector levels, influences on practice and work can include organisational, professional, industrial, cultural, practice and employer expectations. These can be both enablers for, and barriers to, different work practices (see Hall, 2005; Harris & Lyon, 2013).

HUMAN CAPABILITIES AND POSSIBILITIES: A TRANSFORMATIVE VIEW OF THE EDUCATION/PRACTICE NEXUS

We can think of the education/practice nexus as a space where the practice of work meets the practice of education and where transformation can occur, resulting from reflecting, learning and developing capabilities derived from practice engagement. We support the notion of the primacy of practice, meaning that practice gives rise to theory and, following Heidegger, that "everyday practices are the source of intelligibility" (Dreyfus & Hall, 1992, p. 3). Within the practices of education and professional/occupational practice the actions and experience of practice provide the basis for five key outcomes: learning, practice knowledge generation, practice capability development, ethical practice and self-management. Each of these are transformative experiences that provide the basis for ongoing critical self-appraisal, professional development and employability in an age of future uncertainty.

Walker (2012) suggests that a human capabilities model is most adequate to transform and expand higher education. This approach focuses on the intrinsic worth of individuals and seeks to understand what people are actually able to do and be, as opposed to focusing on what resources they have or can generate. The human capabilities approach questions what growth is for, and how education can decrease injustice. This view is in strong contrast to key aspects of the educational and labour market place approach to practice, work and formal education where the notion of commodification of education and professional services is dominant. A human capabilities model of education policy arises from the work of Amartya Sen which views human lives as concerned with the goal of development, not the economy, and focuses on supporting capabilities (Walker, 2012). Globally, education policy is currently dominated by a human capital narrative which frames education as a means to economic growth (Keeley, 2007; Walker, 2012). Human capital is centred on an individual's ability to produce goods and services, where consumption is the primary objective of economic pursuits (Chiappero-Martinetti & Sabadash, 2010). The assumption is that educated workers are better able to generate resources and income. Craft (2013) suggests that if educators conceptualise people as capable and potent, rather than vulnerable and at risk, educators and students, together, can design empowering educational futures. She sees young people as being personally embedded as self-creating consumers and producers within the global capitalist economy which has advantages for those who can use such opportunities positively. However, she regards as problematic the marketisation of young people's possibility thinking that is influenced by the commercial online environment, unless they derive a sense of meaning, identity and life direction both through real and virtual experiences in local and global contexts.

On a positive note, Craft (2013) suggests that co-constructing with young people through the four "Ps" of digital childhood – plurality, participation, playfulness and possibilities – can support creativity that is wise and humanising creativity. Such education futures use collective action to support shared ownership, group identity and empathy that is ethically grounded rather than framed by the marketplace; they create opportunities for meaning making and personal journeys within the collective. Craft (2013) suggests that *what is* is influenced by marketisation, creating individualised narratives for young people, childhood, society and education. In this way, these education futures challenge *what is* in order to imagine *what might be*. It is through the capacities of young people that educators can create with them both their education futures and reconstructed ways of doing and being that are separate from marketisation.

Thus alongside the development of technical occupational capabilities, it is essential that learners realise the importance of ethical and moral conduct in their dealings with others, of critical self-appraisal in their self-management and of taking responsibility to be part of their future world and work arenas as responsible citizens. A key goal (see Chapter 6) is "to help students learn to question the taken-for-granteds in society and how this links to sustainable futures and responsibilities". At the education–practice nexus (often most visible in workplace learning aspects of curricula), students are taking key steps towards becoming capable and self-managed future employable workers and responsible, ethical citizens (whether they are becoming professional practitioners or graduates working outside the established professions).

MEGAN'S VIEW FROM ONTARIO

This section analyses these topics from within the Canadian context; it is particularly situated within the province of Ontario. From this geographical grounding point I (Megan) examine the external and internal socio-political context of future practice and identify possible supports to mitigate these factors and balance the needs of individuals and the needs of the economy. Ontario is the most populous province within Canada. It is defined by an urban corridor skimming the border with the United States and vast rural and remote areas with a declining population base and corresponding declining enrolments in school systems. This has led to a picture of varied and diverse economic growth with certain urban centres producing more robust economic growth and low unemployment rates while rural areas, with a decline in the resource-based economy and low density, demonstrate economic challenges (Freshwater, 2017). Although the following discussion is situated within the context of Ontario, the analysis extends beyond this geographic frame and has applications to other western countries considering how best to build policy and practice models that consider future meta scenarios while remaining adaptive and aware of individual needs. In this section, I focus on education and practice through work integrated learning (WIL) which includes education-enabled cooperative education, internships, field placements and practicums. WIL includes a broad range of practice-work opportunities where learning has a direct work-engagement experience. Practice is, in essence, the thread that links education to work – it represents a way of testing modes of thinking, and ways of doing, and learning a craft. What we as a culture, and as communities, expect about practice reveals a good deal about our values and norms as well as our short- and long-term vision for our desired individual and collective identities.

Within Ontario, there has been a marked increase in rhetoric and policy discourse related to preparing the economy of tomorrow through developing the workforce of today. This policy frame drove a provincial government mandate over the last decade that supported and encouraged WIL for post-secondary learners. WIL is, within the Ontario context, synonymous with "practice work". Responding to this policy initiative, many tertiary educational bodies mandated that all students participate in WIL as part of their education. This direction posed a number of practical and philosophical challenges, especially related to risk management and the universality of such an approach for students, as well as considerations as to how best involve employers and workplace communities in decision making.

The theoretical roots of WIL, as summarised by the Higher Education Quality Council of Ontario (Sattler, 2011), include experiential learning theory (Kolb & Kolb, 2005), situated learning theory (Keating, 2006), activity theory and boundary crossing (Guile & Griffiths, 2001), pedagogy of the workplace (Billett, 2009) and critical education theory (Myers-Lipton, 1998). According to the Higher Education Quality Council of Ontario, almost half of post-secondary students currently have taken part in some form of WIL by the time they graduate (Sattler & Peters, 2013). Most recently, the former provincial Premier commissioned a report that examined ways of building a highly skilled workforce (The Premier's Highly Skilled Workforce Expert Panel, 2016). This 2016 report identified cogent factors facing Ontario's labour market development, educational institutions and, ultimately, the province's ability to generate a strong economy and workforce. In particular, the report outlined a strategy to support the demands of a "technology-driven knowledge economy". The report emphasised:

> that Ontario employers must join their partners in education, labour, government and elsewhere to actively and creatively address regional and

sectoral needs in the labour market and better integrate underrepresented groups including older workers, new Canadians, Indigenous peoples, and persons with disabilities, in an economy that is being rapidly transformed by both demographic and technological change. (The Premier's Highly Skilled Workforce Expert Panel, 2016, p. 3)

The report identified recommendations to build a highly skilled workforce for the future and a number of socio-political factors influencing pathways to work, for a variety of demographics including young people, older individuals and new Canadians. The report suggested that demographic and technological shifts have the greatest influence on individuals seeking employment or education pathways. It painted a picture of shifting dynamics influencing the ability of the province to develop its workforce in a coordinated and impactful way. While the report failed to engage in a critique of the factors impacting the future of practice education, it helpfully identified these factors and their possible impacts on outcomes for individuals and institutions. It also set the policy backdrop against which its analysis was considered.

What this report set out is a dominant narrative related to the importance of human capital as an economic driver. The following identifies some practical suggestions as to how to balance the needs of individual learners with the needs of workplaces and educational bodies such that there is a more equitable balance between a human capabilities frame and the proposed human capital/marketisation and economic frame. Ultimately, if future WIL is to be positioned in a way that equally acknowledges the importance of employers/workplaces, individuals and educational institutions, then concrete supports for all of these actors will be required.

Practical Considerations: Pivoting towards the Future

This policy emphasis on preparing the workforce for tomorrow has generated a series of practical implications that require more critical analysis and unpacking. In particular, this WIL policy focus presents the following questions:

– How best might post-secondary educational institutions build WIL opportunities that pivot towards a future orientation as compared to the way the world and work exist now? In other words, how might the status quo be challenged towards building the world we want as compared to how things exist?
– What capabilities, mind-sets and ways of doing and being might be explored and built for learners in WIL scenarios that will better prepare all learners for the workplaces of tomorrow?
– How might the diverse needs and experiences of learners be considered in balance with the needs of the economy?

Given the uncertain dimensions of the future of work and the corresponding WIL implications, along with associated external factors shaping individuals, workplaces and educational bodies, how might this uncertainty best be responded to holistically and in a way that reflects the diverse needs of all? The following identifies possible supports for all involved.

Supports for Younger People, Employers and Educational Institutions

Given the diversity of backgrounds and places that learners come from and will continue to come from, individuals will need greater access to preparatory supports enabling them to enter the work of professional practice more seamlessly, equitably and, ultimately, better prepared. Regardless of race, class or cultural background, individual learners need concrete supports that provide them with a broad range of critical thinking and professional skills that will enable them not only to enter practice settings with ease but also to thrive in such settings. There has been a well-recognised absence of individuals from diverse backgrounds in a broad range of professions. Given identified institutional barriers to recognition, recruitment and retention of individuals from diverse backgrounds into higher education, what strategies will address comparable barriers within work and practice-work scenarios (Wilson, DePass, & Bean, 2018)? Many social policies prioritise work-entry rather than a more comprehensive approach that looks at a young person's broad range of abilities – through the lens of the capability approach (Egdell & McQuaid, 2016). Too often, the barriers of class, gender, race and place limit learners' abilities to successfully navigate such practice scenarios.

For individual learners who may come from backgrounds that are different than the hegemonic workplace culture (predominantly middle class and white), the rigid social norms and often unspoken rules present hidden barriers and unintended consequences for WIL outcomes (Duffy et al., 2016). With appropriate supports that are both universally available as well as targeted, these barriers could be better addressed and mitigated. Possible supports include mentorship programs that offer perspectives into careers not otherwise observed or known about. Currently, such programming supports are ad hoc and largely facilitated by the charitable sector. Or, if they are provided, often those who might benefit from mentoring supports are not able to access them or participate meaningfully in these initiatives (Jones, 2017). Similarly, how best to deliver these supports in light of regional and place difference remains a practical challenge to be resolved. Learning and coaching that exposes individuals to professional skills and reinforces workplace-appropriate examples of professionalism in sustained ways will also be helpful. Professionalism cannot be the code-word for "you're different and therefore you are not a professional". Realising that difference will become an increasingly important value in the future, such learning opportunities need to bend consciously and reflectively towards including difference and a critical lens regarding what is professionalism and how it is to be taught.

Workplaces of the future will increasingly need competencies related to complexity, divergent and integrated thinking, as well as high degrees of adaptability and a willingness to take risks and continuously learn. Increasingly, innovation and entrepreneurship are being identified as core competencies for workplaces in the future – however, learning these skills is problematic, especially from a capability approach (Morselli, 2017). Presently, many professional learning programs are structured in such ways as to limit failure and ensure safety at all costs. The challenge for those facilitating WIL will be to continue to challenge the status quo in ways that are within limits of acceptable risk while finding new solutions to emerging problems. Learning boundary-crossing will be a critical competency in the future (Walker & Nocon, 2007). This is especially true in workplaces that address or attempt to tackle complex social, environmental or health problems. For instance, large emerging problems such as climate change or systemic poverty will require new solutions to improve outcomes; however, WIL scenarios remain limited in their ability to expose those participating to divergent ways of solving problems or identifying new solutions.

Possible ways to support this adaptability mindset and ability to utilise divergent thinking include design thinking and supporting complexity and critical thinking within WIL scenarios (Wright & Cairns, 2011). Problem solving and an ability to work in interdisciplinary teams will also be encouraged approaches in the future and these can be built into WIL settings with appropriate and sufficient support for all those involved. Using scenarios, simulated settings and complex interdisciplinary team learning approaches will help enable this competency. Working with technology and possibly virtual reality simulators to identify the types of problems and challenges presented in future work and identifying technological solutions will be helpful to future workplaces; these competencies can be encouraged and supported through WIL.

None of these supports will necessarily guarantee that young people are effectively positioned to achieve the types of outcomes they desire through identified WIL in the future. The horizon of WIL remains constantly in flux and unfolding. Providing as many supports as possible to level the playing field and to build up competencies for individuals will allow them to thrive in a variety of WIL settings remains the goal. The fundamental challenge will be for educational institutions to rethink their pedagogical approaches in ways that learners will be able to make effective choices and to embody mindsets that will allow them to thrive in a changing and uncertain future.

Workplaces and educational bodies are situated in places influenced and shaped strongly by their geography as well as by local and global political, cultural and socioeconomic dimensions (e.g. demography). Helping workplaces and educational institutions better prepare for practice in the future requires a nuanced set of policy and practical supports that recognise the diverse realities each of these structures are situated within. Often the challenge of planning for a future that is unknown is that organisations such as workplaces and educational bodies are both enlivened and constricted by values and norms of what is acceptable and what isn't within their specific context. Challenging these norms or attempting longer-term planning is problematic as it often requires significant resources as well as risk taking. This requires bold leadership and a comfort level with setting longer-term goals and readjusting depending on what is happening on the socioeconomic and political horizon. The challenge posed to workplaces today to become more ready for the workplaces of tomorrow is essentially a challenge of capacity and a readiness to adapt and learn reflectively. Emphasis on readiness for the future cannot be placed on young learners alone. It needs to be shared by educational bodies and workplaces as well as government and policy makers.

Implementing the identified supports will rely largely on governments and policy makers who might incentivise workplaces to support more of an adaptive pedagogy and increased opportunities for WIL for workplaces. Similarly, emphasising the importance of quality rather than quantity in WIL is another essential goal of helping workplaces and educational organisations to prepare themselves for an uncertain future. This requires establishing quality standards and supporting workplaces and educational bodies through building capacity, understanding the importance of standards, and providing knowledge and education about the shifting workplace dynamics.

Much of the future of work may not yet be known or recognisable — all stakeholders will need to navigate the socio-political factors that are emerging within a province like Ontario and recognise the shifting landscape and its impacts on individuals as well as workplaces and employers and the educational bodies involved. Within Ontario some work has been done to begin to craft conversations between employers and educational bodies related to supporting learning for practice in practice settings – this type of collaborative dialogue will enable stronger learning scenarios and greater opportunities moving forward. This has happened largely through government investments in pilot projects to support WIL. Employers and educational bodies have largely led these with limited involvement from youth. However, with a recent change in government, the extent to which support for WIL and associated funding opportunities will remain is unknown. This indeed remains the challenge for all stakeholders now and presumably in the future – what WIL looks like, even what it is called and the supports that might be on offer to facilitate broader engagement with it, may continue to change or look slightly different depending on what is politically supported or encouraged or what is emerging as a dominant trend. This requires capacity and adaptation from stakeholders to address shifting trends in the WIL frame itself and also in the emerging socio-political factors influencing the context.

CONCLUSION

This discussion is situated in a critique of the emerging neo-liberalist rhetoric and corresponding policy frames that dominate the construction of workplaces and education programs as primarily engines for broader economic/structural challenges rather than more emancipatory platforms that centrally position individuals' desire to achieve capacity or build and strengthen capabilities towards longer-term success and pathways for career mobilisation, actualisation, etc. Although it is simplistic to acknowledge singularly the significance of building human capacity and human capabilities – the question remains as to what and who remains peripheral when the economy and human capital are primary drivers. Are institutions more critical than the individuals they serve? Or in other words, who or what sits at the centre of discourses related to education for work? Are institutions, the labour market and, ultimately, economic growth the aim of any policy considerations, or does individual development and actualisation matter more? These are philosophical questions that present themselves and require reconciliation. In this chapter we propose a blended approach that acknowledges both the importance of the individual and the importance of the economy in advancing integrated policy decisions and work recommendations. The individual cannot be at the forefront without consideration for the economy nor can the economy be primary without consideration for the individual.

Currently, the pathway(s) for young people towards future work in practice remains contested and constrained by competing policy, cultural and socioeconomic factors. However, youth are not solely the casual/passive recipients of external factors – with appropriate supports they hold capacity to make powerful choices. Similarly, there are alternate learning pathways for young people that might be more inclusive and reflexive of socioeconomic and political factors. This requires a solid rethinking of the pedagogical approaches that inform education as well as how best to enable practice-based education (particularly WIL) in ways that align to the complexity of what might be present in the future.

Governments, employers and educational bodies alongside young people may be able to co-design alternate pathways that identify stopping points for young people from early education through to career pathways using scenario planning and design thinking methodologies. Creating these pathways and building them together might illuminate ways of generating practice futures that are inclusive and better positioned for the needs of all stakeholders. A readiness to bend and adapt both practice and learning strategies at the various intersections of these practice pathways will be the real challenge for all stakeholders involved.

REFERENCES

Altbach, P. G. (1992). Higher education, democracy and development: Implications for newly industrialized countries. *Interchange, 23*(1&2), 143-163.

Barber, B. R. (1995). *Jihad vs. McWorld*. New York, NY: Ballantine Books.

Baston C. D. (1994). Why act for the public good? Four answers. *Personality and Social Psychology Bulletin, 20*(5), 603-610.

Biddle, N., Sheppard, J., & Gray, M. (2018, July 31). Australians worry more about losing jobs overseas than to robots. *The Conversation*. Retrieved from https://theconversation.com/australians-worry-more-about-losing-jobs-overseas-than-to-robots-100728

Billett, S. (2009). *Developing agentic professionals through practice-based pedagogies* (Final Report for ALTC Associate Fellowship). Retrieved from https://altf.org/wp-content/uploads/2016/08/Billett_S_Associate-Fellowship-Final-Report.pdf

Boix, C., & Posner D. N. (1988). Social capital: Explaining its origins and effects on government performance. *British Journal of Political Science, 28*(4), 686-693.

Brown, R., Copeland, W. E., Costello, E. J., Erkanli, A., & Worthman, C. M. (2009). Family and community influences on educational outcomes among Appalachian youth. *Journal of Community Psychology, 37*(7), 795-808.

Chatty, D. (2010). *Displacement and dispossession in the Modern Middle East*. New York, NY: Cambridge University Press.

Chiappero-Martinetti, E., & Sabadash, A. (2010). *Human capital and human capabilities: Towards a theoretical integration* (Paper prepared for the workshop 'Workable'). Pavia: University of Pavia.

Craft, A. (2013). Childhood, possibility thinking and wise, humanising educational futures. *International Journal of Educational Research, 61*, 126-134.

Davis, S. (2018, January 11). Soft robots could be the factory workers of the future. *The Conversation*. Retrieved from https://theconversation.com/soft-robots-could-be-the-factory-workers-of-the-future-89885

D'Olympio, L. (2015, July 15). Caring about climate change: Global citizens and moral decision making. *The Conversation*. Retrieved from https://theconversation.com/caring-about-climate-change-global-citizens-and-moral-decision-making-44771

Dreyfus, H., & Hall, H. (1992). Introduction. In H. L. Dreyfus & H. Hall (Eds.), *Heidegger: A critical reader* (pp. 1-25). Oxford, England: Blackwell.

Duffy, R. D., Blustein, D. L., Diemer, M. A., & Autin, K. L. (2016). The psychology of working theory. *Journal of Counseling Psychology, 63*(2), 127-148.

Duffy-Jones, R., & Argent, N. (2018, August 8). Why young women say no to rural Australia. *The Conversation*. Retrieved from https://theconversation.com/why-young-women-say-no-to-rural-australia-100760

Egdell, V., & McQuaid, R. (2016). Supporting disadvantaged young people into work: Insights from the capability approach. *Social Policy & Administration, 50*(1), 1-18.

Epitropaki, O., & Martin, R. (1999). The impact of relational demography on the quality of leader–member exchanges and employees' work attitudes and well-being. *Journal of Occupational and Organizational Psychology, 72*, 237-240.

Ferris, E., & Kirişci, K. (2016). *Consequences of chaos: Syria's humanitarian crisis and the failure to protect*. Washington, DC: The Brookings Institution.

Flippen, A. (2014, June 26). Where are the hardest places to live in the U.S.? *The New York Times*. Retrieved from https://www.nytimes.com/2014/06/26/upshot/where-are-the-hardest-places-to-live-in-the-us.html?_r=0&abt=0002&abg=0

Florida, R. (2018, September 18). The divides within, and between, urban and rural America. *Citylab*. Retrieved from https://www.citylab.com/life/2018/09/the-divides-within-and-between-urban-and-rural-america/569749/

Freshwater, D. (2017). *Rural Ontario Foresight Papers: Growth beyond cities*. Rural Ontario Institute. Retrieved from http://www.ruralontarioinstitute.ca/uploads/userfiles/files/Rural%20Ontario%20Foresight%20Papers%202017_Growth%20beyond%20cities%20and%20northern%20perspective.pdf

Gambles, R., Lewis, S., & Rapoport, R. (2006). *The myth of work-life balance: The challenge of our time for men, women and societies*. West Sussex, England; John Wiley & Sons.

Gonzalez, J. A., & Denisi, A.S. (2009). Cross-level effects of demography and diversity climate on organizational attachment and firm effectiveness. *Journal of Organizational Behavior, 30*, 21-40.

Guile, D., & Griffiths, T. (2001). Learning through work experience. *Journal of Education and Work, 14*(1), 113-131.

Hall, P. (2005). Interprofessional teamwork: Professional cultures as barriers. *Journal of Interprofessional Care, Supplement 1*, 188-196.

Hancock, P. (2018, January 24). Don't automate the fun out of life. *The Conversation*. Retrieved from https://theconversation.com/dont-automate-the-fun-out-of-life-88681

Harris, F., & Lyon, F. (2013). Transdisciplinary environmental research: Building trust across professional cultures. *Environmental Science & Policy, 31*, 109-119.

Held, D., & Hale, T. (2017, November 8). The world is in economic, political and environmental gridlock – here's why. *The Conversation*. Retrieved from https://theconversation.com/the-world-is-in-economic-political-and-environmental-gridlock-heres-why-85641

Heller, D. E., & Callender, C. (Eds.). (2013). *Student financing of higher education: A comparative perspective*. Abingdon, England: Routledge.

Howcroft, D., & Rubery, J. (2018, June 11). Automation has the potential to improve gender equality at work. *The Conversation*. Retrieved from https://theconversation.com/automation-has-the-potential-to-improve-gender-equality-at-work-96807

Humphreys, M. (2003). *Economics and violent conflict*. Cambridge, MA: Harvard University. Retrieved from https://www.unicef.org/socialpolicy/files/Economics_and_Violent_Conflict.pdf

Jackson, D., Rowbottom, D., Ferns, S., & McLaren, D. (2016). Employer understanding of work-integrated learning and the challenges of engaging in work placement opportunities. *Studies in Continuing Education*, 1-17.

Jones, M. (2017, June 2). Why can't companies get mentorship programs right? And when they get them wrong, it can be worse than having not tried at all. *The Atlantic*. Retrieved from https://www.theatlantic.com/business/archive/2017/06/corporate-mentorship-programs/528927/

Kak, S. (2018, January 9). Universities must prepare for a technology-enabled future. *The Conversation*. Retrieved from http://theconversation.com/universities-must-prepare-for-a-technology-enabled-future-89354

Kalbfleisch, P. J. (2003). An organizational communication challenge to the discourse of work and family research: From problematics to empowerment. In P. J. Kalbfleisch (Ed.), *Communication Yearbook 27* (pp. 1-44). New York, NY: Routledge.

Keating, J. (2006). Post-school articulation in Australia: A case of unresolved tensions. *Journal of Further and Higher Education, 30*(1), 59-74.

Keeley, B. (2007). *Human capital: How what you know shapes your life*. Paris, France: OECD.

Kolb, A. Y., & Kolb, D. A. (2005). Learning styles and learning spaces: Enhancing experiential learning in higher education. *Academy of Management Learning & Education, 4*(2), 193-212.

Malatest, R. A., & Associates Ltd. (2002). *Rural youth study, Phase II – Rural youth migration: Exploring the reality behind the myths* (A Rural Youth Discussion Paper). Government of Canada. Retrieved from http://www.ruralontarioinstitute.ca/file.aspx?id=dfa7c6b8-ae0a-4366-9439-791b7db17dbc

Mansfield, E. D., & Pollins, B. M. (Eds.). (2003). *Economic interdependence and international conflict: New perspectives on an enduring debate*. Ann Arbor, MI: The University of Michigan Press.

Mayer, K. U. (2004). Whose lives? How history, societies, and institutions define and shape life courses. *Research in Human Development, 1*(3), 161-187.

Minku, L. L., & Levesley, J. (2018, August 20). AI doctors and engineers are coming – but they won't be stealing high-skill jobs. *The Conversation*. Retrieved from https://theconversation.com/ai-doctors-and-engineers-are-coming-but-they-wont-be-stealing-high-skill-jobs-101701

Morselli, D. (2017). Boundary crossing workshops for enterprise education: A capability approach. In P. Jones, G. Maas, & L. Pittaway (Eds.), *Entrepreneurship education: New perspectives on research, policy and practice* (pp. 277-300). Bingley, England: Emerald.

Myers-Lipton, S. (1998). Effect of a comprehensive service-learning program on college students' level of civic responsibility. *Teaching Sociology, 26*(4), 234-258.

Nye, J. S. (2004). *Power in the global information age: From realism to globalization*. London, England: Routledge.

The Premier's Highly Skilled Workforce Expert Panel. (2016). *Building the workforce of tomorrow: A shared responsibility* (Report submitted to the Honourable Kathleen Wynne, Premier of Ontario). Retrieved from https://www.ontario.ca/page/building-workforce-tomorrow-shared-responsibility

Riordan, G. M., & Shore, L. M. (1999). Demographic diversity and employee attitudes: An empirical examination of relational demography within work units. *Journal of Applied Psychology, 82*(3), 342-358.

Rosenau, J. N., & Singh J. P. (2002). *Information technologies and global politics: The changing scope of power and governance*. New York, NY: SUNY Press.

Sattler, P. (2011). *Work-integrated learning in Ontario's postsecondary sector*. Toronto, Canada: Higher Education Quality Council of Ontario.

Sattler, P., & Peters, J. (2013). *Work-integrated learning in Ontario's postsecondary sector: The experience of Ontario graduates*. Toronto, Canada: Higher Education Quality Council of Ontario.

Schwab, K. (2017). *The Fourth Industrial Revolution*. London, England: Penguin Books.

Smith, M. E. (2011). A liberal grand strategy in a realist world? Power, purpose, and the EU's changing global role. *Journal of European Public Policy, 18*(2), 144-163.

Walker, D., & Nocon, H. (2007). Boundary-crossing competence: Theoretical considerations and educational design. *Mind, Culture, and Activity, 14*(3), 178-195.
Walker, M. (2012). A capital or capabilities education narrative in a world of staggering inequalities? *International Journal of Educational Development, 32*(3), 384-393.
Warschauer, M. (2000). The changing global economy and the future of English teaching. *TESOL Quarterly, 34*(3), 511-535.
Wilson, M. A., DePass, A. L., & Bean, A. J. (2018). Institutional interventions that remove barriers to recruit and retain diverse biomedical PhD students. *CBE—Life Sciences Education, 17*(2), Art. 27.
Wright, G., & Cairns, G. (2011). *Scenario thinking: Practical approaches to the future*. Basingstoke, England: Palgrave-Macmillan.

Megan Conway PhD (ORCID: https://orcid.org/0000-0001-6147-315X)
Chair, Health and Community Studies
Algonquin College in the Ottawa Valley, Canada

Joy Higgs AM, PhD (ORCID: https://orcid.org/0000-0002-8545-1016)
Emeritus Professor, Charles Sturt University, Australia
Director, Education, Practice and Employability Network, Australia

PART 3

PURSUING PRACTICE FUTURES

FRANZISKA TREDE AND JOY HIGGS

11. THE PLACE OF AGENCY AND RELATED CAPACITIES IN FUTURE PRACTICES

In this book on practice futures a number of the major themes deal with larger dimensions of system and cultural changes. The future of the professions and professionalism is a topic examined in Chapters 6 and 7. In this chapter we focus on the people, especially professionals as human agents, in this contested space and explore the place agency in shaping future practices.

Agency is the capacity to act and not feel helpless. It leads to purposeful and responsible action that is distinguished from following rules and procedures without reflection. Furthermore, agency enhances independence and responsibility in professional decision making with clients, and when focused on decreasing social inequalities, contributes to the betterment of society.

The way we understand agency, as a unique human characteristic, has always been contested, to a degree, and has come again into focus with the advances of artificial intelligence. Although sociocultural and material contexts have always been acknowledged as influencing agency, the role of non-human agency is finding currency. Knappett and Malafouris (2008), with their edited book on material agency, propose a rethinking of agency away from an overly human-centred approach. With this chapter, we explore three main concepts: liquid times, the Fourth Industrial Revolution and the place of human and non-human agency in future practices. We start by exploring different theoretical perspectives on framing agency. We focus on four heuristics – essentialist, pragmatic, identity, and career wide – and carve out agency within a temporal frame. We then discuss the interdependent relationship of agency with identity, resilience and capabilities. Next we introduce the concept of the *deliberate professional*, a term coined by Trede and McEwen (2016) and propose this idea as a useful framework for understanding the place of agency in future practice.

The deliberate professional is thoughtful yet assertive and action-oriented, considers possibilities and reflects on other ways of acting to improve conditions and practices for self, others and the common good. From this discussion we conclude that human agency – regarded in a complex, interdependent relationship with sociocultural structures and other actors in the practice arena – plays a crucial role in enabling professionals to practise effectively for the good of people and society, in future practice.

THE CHANGING NATURE OF WORK, CAPABILITIES AND THE HUMAN CONDITION

When we set out to imagine what practice might be like in the future we are aware that we cannot be certain, but we can glean an understanding of where we are heading as the world is becoming more connected, diverse and complex. There are core future conditions and developments on the horizon that will play into the way practices will evolve. These include the increase of automation, sophistication of artificial intelligence, and a changing economy with increasing inequality. These developments mean that the nature of work and the workplace will change as well.

Changes in technology, globalisation and the economy have already increased in speed and pervasiveness. They have had a much more profound impact on personal, professional and public lives than any previous economic revolution (Fuchs, 2016). Bauman (2012) coined the term *liquid times*, meaning that the old ways of doing things no longer fit the new conditions. We are in transition. The two key points he emphasises are time and uncertainty. We experience relentlessly rapid and constant change which does not allow time for pausing and reflecting. Standing still actually means falling behind. Structures, strategic plans and practices liquify before they have time to solidify.

And secondly, Bauman points out that the human condition is such that we have given up on control, predictability and certainty, and acknowledge that we live in uncertain, fragile conditions. In liquid times (or the Fourth Industrial Revolution or liquid modernity) we are exposed to ambiguity, plurality and possibilities. Chaotic phenomena that have emerged include fake news and the gig economy (where organisations contract with independent workers for short-term engagements). In these times supposedly everything is possible but there is no assurance of best outcomes for all. Agency is a key to successful outcome; it can be conceptualised as a disposition for courageous and purposeful practice, and it promises to help navigate professionals into future practices.

With advances of technology new possibilities are opening up that question the way we practise, work with technology and relate to each other. Machines can do routine work processes with greater precision and infinite patience; both of these attributes can lead to increased productivity and improved client experiences. For example, robotic policemen can stoically and patiently manage traffic without taking a social position. With advances in algorithmic and digital techniques, machines can produce graphs and statistics from available data, outperforming any analysis by humans. These technical capacities then free up time and even necessitate professionals to rethink the human skills required in the future to provide professional services to clients.

Future capabilities of humans whose roles are not overtaken by artificial intelligence will focus on those that are unique to human beings such as empathy, creativity and judgement. This changing nature of work and future practices raises issues around the interface between humans and machines; for example, how do we effectively employ human capabilities and productively engage with and work alongside technology? Keeping up with these profound shifts in future practice capabilities requires human agency.

FRAMING AGENCY

Traditionally, agency has been thought of as a human-centred capacity that resides in individuals. To be human means having agency, recognising that this is a variable influenced by both personal abilities and external constraints and opportunities. Some people have very high levels of agency, others have little. Agency is the capacity to act with consciousness and purpose, and to have some control and influence over one's life.

There are different theoretical stances to interpret this framing of agency. For example, to what extent individuals can act freely remains contested. Viewed from social sciences perspectives, the capacity to act is socioculturally mediated (Ahern, 2001) and depends on others, objects, plus time and place. The will to act and the belief that one can act is understood as not purely individualistic because individuals are constrained to act within social structures and relationships. That is, agency is socially constructed. There are people, typically those with greater power, who support or hinder other people's actions. In social sciences the ongoing debate about agency versus structure (Archer, 1988; Bourdieu, 1972/1977; Giddens, 1984) crystallises the tensions between individual actions and established cultural conventions. Archer privileges agency over structure whereas Bourdieu has a slightly more pessimistic perspective on human agency and attributes structures with more prominence.

This debate extends into the discourses of practice theory (Bourdieu, 1990). Practices and their social structures already exist within fields of practice and are often unconsciously reproduced by practitioners. Practice occurs in cultural settings with established norms, taken-for-granted work routines and shared expectations. However, practice is not only reproduced; practice is dynamic and changing. Practice occurs in fields of competing interests and there are choices and possibilities to act otherwise.

Agency, then, is a product of engaging with these fields of practice. Broadly speaking, theorists who have engaged in the agency versus structure debate agree that this binary is untenable and needs to be understood as an interdependent relationship. Agency is a mediating activity that requires engagement of individuals with the social and material world. This broad framing of agency helps to locate agency within the social arena but is of little help in defining or identifying agentic action. Here we need to pay closer attention to the self. Bandura (2001) has put forward intentionality, forethought, self-regulation and self-efficacy as four aspects of individual agency. These are agency processes that occur before individuals act.

The processes of enacting agency include identifying options, making choices, exercising judgement and taking responsibility for actions. Depending on context, the degree of agency enacted by an individual can vary from very little and drifting with change, to stronger agency and making changes or even to actively resisting current practices.

What it means to have agency depends on the purpose of enabling action. The purpose of agency can relate to making changes in personal, professional or public life. Each of these require intentional engagement with different contexts.

Individuals with personal agency have a perceived sphere of influence on (and by) family and friends; individuals with professional agency have a perceived sphere of influence on working with colleagues, professional bodies and clients – and individuals with public agency have a perceived sphere of influence on broader social and political arenas. Another aspect of agency is time.

Agency can have differential orientations ranging from short-term or long-term purposes. In an acute crisis situation immediate action is required whereas fulfilling one's life dream will require a longer and committed pathway. Agency then can be understood to relate closely with time and situatedness. Hitlin and Elder (2007) have developed a useful framework to think about agency as self in relation to time. They suggest analysing agency through a temporal lens. "A temporally based heuristic offers a schematic for understanding multiple, sometimes conflicting, uses of agency" (ibid, 2007, p. 175). Their temporal framework helps to mediate ideas of sense of control, free will and structural contexts. The four types of agency they put forward are existential, pragmatic, identity and life-course agency, as discussed below.

Existential Agency

Within social constraints humans are free to act. They have choices in how to behave and options in how to act otherwise (Giddens, 1984). This existential agency, however, is dependent on self-perceptions about the ability and power to act in one's social world. An engineer might have the social and professional standing to act differently but might not have the self-belief to do so. Existential agency is as much about self-belief as it is about enabling social structures (Bandura, 2001).

Pragmatic Agency

A person's pragmatic agency concerns immediate situations that involves oneself with the social world. It relates to routine practices and when these routines break down (Dewey, 1933). Pragmatic agency is the ability to innovate. As such, pragmatic agency relates to creativity and emergent actions that are not purely reactions to situations. Pragmatic agency comes into play in problematic, critical moments when routine actions are no longer an appropriate solution.

Reflective, emotional as well as dispositional elements inform pragmatic agency. It deals with actions that relate closely to the social environment. When the concept of self no longer corresponds well with cultural expectations something needs to shift to readjust expectations. Emotions play a part in this type of agency that entangles self with the social.

Identity Agency

Identity agency represents the idea that we develop self-identity within social norms. We are not all identical and we act differently although we live within shared social norms and expectations. Identity agency is about enacting one's role, be it in the family, at work or in public life. Identity agency is about "achieving desired social or substantive ends" (Hitlin & Elder, 2007, p. 180). People play roles of a teacher or a parent (and so on) and are acting according to social expectations, living up to commitments and routine pathways.

Receiving affirmative feedback about actions taken strengthens identity agency and a sense of belonging to a community. Positive feedback motivates reproduction of practices and structures (Bourdieu & Giddens, 1984). An amplified self-commitment to identity agency can reinforce a taken-for-granted attitude towards habitual or chosen actions. Tensions between "keeping face" and remaining credible within socially binding structures and acting according to one's own values collide in identity agency. Self-identity is interdependent within a sociocultural space just like professional identity is shaped and shapes the world of professions. Identity agency is about enacting internalised values.

Life Course Agency

Life course agency is the belief in one's ability to influence one's life journey in a longer time frame. Actions that are not a response to an immediate situation with short-term outcomes but are aimed at achieving future goals define this type of agency. The reflexive belief and the ability to interpret situations within longer-term time frames relates closely to being able to see options and having a sense of control over one's life.

Who we want to become signifies life course agency. It is used at significant turning points in people's lives. People's life journeys and careers are steered by abilities to plan ahead combined with self-belief, and of course, life course agency is also always embedded within histories (life milestones) and within social structural constraints.

Reflection on Agencies

These four types of agencies seen through a temporal perspective are helpful in understanding the role of agency for future practice as they put into perspective individuals' current situations as well as their life-wide career journeys. Agency relates to existential, situated and long-term aspects of people's lives, behaviours and actions. All four types are closely interwoven and overlap. Existential and life course agency are the foundational types that provide a broader frame for agency whereas pragmatic and identity agency relate more closely to immediate life situations. Life course agency relates closely to ideas of identity, resilience, capability and intelligences which we will discuss next.

CULTURE, SOCIALISATION AND IDENTITY

From the previous section it is evident that a person's agency is shaped by their temporal horizons (where they are up to in their life journey), their values, experiences, choices and interests as well as the situations (cultures, practice worlds, workplaces, etc.) they encounter and inhabit. In this section we discuss the two key dimensions of culture and socialisation and how this links with identity, specifically in relation to practice/professional culture/socialisation.

Different occupations create their own occupational and workplace cultures that underpin how people work together. Deal and Kennedy (1982) defined culture in relation to "the way we do things here" and Schein (2004) identified three levels of culture: artefacts and behaviours (outward and visible elements of the setting), espoused values (stated values of the organisation/group) and assumptions (deeply held shared assumptions of the organisation/group). Understanding the culture of occupations and workplaces is a key to preparing programs of occupational socialisation and identify formation.

In these pursuits we need to remember that culture is not entirely visible or explicit. Hall (1977) recognised two "ice-berg-like" aspects of culture: externally conscious dimensions (e.g. behaviours) and internally conscious dimensions that are below the surface (e.g. values, motivations, prejudices, attitudes). Looking at these views of culture we recognise the importance of acculturation or socialisation by novices as they enter occupations, teams and workplaces.

> Professional socialisation could be thought of as the individual's journey in becoming a member of a particular profession, a unique social group, and learning to be part of the culture of that group with all its privileges, requirements and responsibilities. (Higgs, 2013, p. 86)

> During professional socialisation novices encounter dialogues and experiences which challenge, extend or affirm their entry values and interests or motivations. Topics such as social responsibility, professional ethics and codes of professional conduct typically form part of professional education curricula and are faced personally during workplace learning experiences. (ibid, p. 85)

Through socialisation novices gain a range of occupationally relevant knowledge, values, understanding of norms and expectations, behaviours and abilities that enable them to belong compatibly and work relevantly in that occupation and its various practice communities. This transformation relates to Bourdieu's (1972/1977) idea of habitus as the product of socialisation. Habitus is described by Wacquant (2005) as "the way society becomes deposited in persons in the form of lasting dispositions, or trained capacities and structured propensities to think, feel and act in determinant ways, which then guide them" (p. 316).

"By sharing cultural norms, knowledge and practices, members of (practice) communities function coherently as members of a group who walk, talk and think (reason) in shared, encultured ways" (Higgs, 2019, p. 17). In this way novices and more experienced practitioners develop and evolve their occupational identities.

Various researchers have explored different aspects of professional identity development. Reid et al. (2008) contend that "professional identity formation is a relation between students' learning experience and the manner in which they anticipate or practise in professional working life" (p. 729). Henschke (2012) drew attention to the particular benefit of cooperative education or work integrated learning as a valuable means of exploring and developing multiple identities: **I** (a sense of self as an aspiring professional, self-awareness and self-belief), **You** (a sense of how others see you and how people develop relationships and engage in social interactions and work practices) and **Us** (being part of a practice community).

Further to these ideas, Fellenz (2015) studied the formation of the professional self in the context of professional practice requirements and proposed an ontological (being and becoming) perspective. He raises the question of the autonomy paradox: what type of graduate is education producing in relation to agency amidst the competing influences on the formation of the professional self, of the sovereign self and the external authority of the profession. He argues that the notion of *bildung* can help understand this matter.

> as the concept of Bildung highlights the duality of (internal) self formation processes and of (external) relationships between the self and its social, political, technological and cultural context, it can help to consider how the formed professional self can retain enough autonomy to challenge professional orthodoxy, for example, in situations where unique circumstances contribute to the failure of standard practice to deliver intended outcomes; in the light of new evidence arising from practice that challenges canonical professional approaches; when new technological developments or other sociomaterial affordances create opportunities not covered by previous collective professional consideration; or in the context of ethical dilemmas that require unorthodox responses. (Fellenz, 2015, p. 13)

This identity development through external as well as internal development was taken up by Cornelissen and van Wyk (2007), who recognised the role of the learning and external influences in identity development. To them there are three core elements of the learner's place in role development and commitment to the professional role: knowledge acquisition, investment and involvement. Similarly, De Weerdt et al. (2006) stress the importance of identity (attitudes, values and beliefs) development and transformation and whole person development, alongside the development of knowledge and skills in professionals and "knowledge" industries, through "deep" professional development. The key to this deep process is seeing identity transformation as an intercontextual process that has two connected sub-processes of development where a) identity transformation is content-related (i.e. the story of who we are and also what we stand for), and b) identity transformation is process-related (i.e. our sense of self-worth and self-confidence is developed and gives rise to inner safety and the capacity to drive self-transformation). They see identity development as both an individual and relational process.

Practitioners, as discussed above, participate in their own professional identity development. We can expect them to vary in the extent to which their identity relates to their occupation's practice ranging from the centre, the mainstream, to the practice fringes, to spaces of practice contestation. These positions may have been shaped by hegemonic influences (including formal education and workplace norms) and rewards (including career progression and job satisfaction). They may be managed by choice, particularly in highly agentic practitioners and deliberate professionals who choose to shape and pursue "their practices within a coherent and deliberately owned practice model" (Higgs, 2016, p. 189).

AGENCY AND RESILIENCE, CAPABILITIES AND INTELLIGENCES

Agency does not occur in isolation. People need capabilities that support actions and justify decisions and actions of agents. People need emotional intelligences to guide decision making and deal with the impact of agential actions on themselves and others. People need resilience to deal with the problematics and complexities of endeavouring to purse agency when constrained by a range of factors (e.g. policies, resources, power) that influence and control the possibilities and their actions.

Many words are used to label the idea of what people bring to their work, learning and practice that enable them to complete these tasks and roles: knowledge, learning, attitudes, dispositions, attributes, competencies, abilities and capabilities. In terms of the latter four of these we see a range of levels of precision, measurability, versatility and capacity.

Here we argue, among these, for the use of *capability* since it best addresses the demands of complexity and unpredictability in practice. According to Stephenson (1998, p. 1):

> Capability is an all round human quality observable in what Sir Toby Weaver describes as 'purposive and sensible' action (Weaver, 1994). Capability is an integration of knowledge, skills, personal qualities and understanding used appropriately and effectively – not just in familiar and highly focused specialist contexts but in response to new and changing circumstances. Capability can be observed when we see people with justified confidence in their ability to:
>
> – take effective and appropriate action;
> – explain what they are about;
> – live and work effectively with others; and
> – continue to learn from their experiences as individuals and in association with others, in a diverse and changing society (Stephenson, 1992).

In this portrayal of capability we see both the foundation or ability that underpins agency as well as an overlap with key dimensions of agency (i.e. taking and justifying appropriate action). Similarly, the idea of agency is embedded within capability. They are symbiotic actions, potentials and abilities.

Neugenbauer and Evans-Brain (2016) remind us that intellectual intelligence by itself does not predict success in work. They argue that underlying emotional intelligence is the assumption that people who can recognise their own emotions and those of others, and who can understand social situations and manage their own emotions, will be better placed to achieve success in work and life situations.

In practice, in general and particularly in relation to agency, emotional intelligences are important in such success (see Mayer, Caruso, & Salovey, 2016).

> Emotional intelligence refers to an ability to recognize the meanings of emotion and their relationships, and to reason and problem-solve on the basis of them. Emotional intelligence is involved in the capacity to perceive emotions, assimilate emotion-related feelings, understand the information of those emotions, and manage them. (Mayer, Caruso, & Salovey, 1999, p. 267)

Resilience can be thought of as a component of emotional intelligence. "In the workplace it is about the ability to adapt and to recover or bounce back when faced with change or adversity" (Neugenbauer & Evans-Brain, 2016, p. 141). Youssef and Luthans (2007) differentiate between resilience and other positive attributes such as hope and optimism; resilience can involve both being proactive and reactive in the face of adversity.

Through these interpretations of resilience and emotional intelligence we see a valuable link to agency. The agent, being purposeful and seeking to retain control and take responsibility for their choices and actions in both positive and negative situations, needs to have strategies that effectively use emotional intelligences and resilience. In relation to being an agent that leads initiatives and groups it is useful to reflect on the work of George (2000), who proposed that emotional intelligence contributes to effective leadership in relation to five essential elements of leadership:

- development of collective goals
- helping others appreciate the importance of work activities
- generating and maintaining enthusiasm, optimism, confidence, trust and cooperation
- encouraging flexibility in decision making and change
- establishing and maintaining a meaningful identity for an organisation.

Returning to the chapter title and its reference to future practice we propose that the future will demand self-knowledge and self-managed identity. It will also require flexibility and the capacity to respond to the unknown, and preparation to work within chosen practice communities and workplaces. Workers will need the capability to adjust and learn to be able to change career direction and work in new and unpredictable ways and spaces. Finally, they will need durable attributes like resilience and sustainability along with the capacity to develop goals and strategies to take these strengths into innovative practices. Across each of these lie the need for self-critique, self-confidence and agency.

TOWARDS THE DELIBERATE PROFESSIONAL

Building on resilience, capabilities and intelligences in the previous section, we now turn to explore the role of purpose and choice in developing a sense of agency and acting in the social world. Agency alone has little meaning; the capacity to act requires a purpose: why and what for? Furthermore, the capacity to act and not feel helpless relies also on having choices regarding how to act. Through awareness of, and engagement with, power differentials, privilege and social justice issues we can come to better understand the histories of current practices and how they came into being. This deeper understanding of how current practices became the status quo will help us to think about possibilities and choices in how to transform them into the future.

Being curious about how practices keep evolving heightens awareness of hierarchies, power games and capitals (Bourdieu, 1990). Mixed with courage this curiosity or critique opens up possibilities for how future practices could be imagined and shaped differently. Social structures, social relationships and individuals are constructed by humans and therefore are dynamic rather than static and homogeneous. Indeed if practices never changed all we would need to do is learn and memorise routine procedures. Agency, and with it, choice and purpose, would get in the way of perpetuating practices. So in the spirit of co-creating the future of practices, individuals need to cultivate a sense of deliberateness and agency.

The degree to which people have agency differs; this is not just because of themselves but also due to their race, gender and cultural norms (hooks, 1990). For example, in a male dominated profession such as engineering, a woman engineer has her agency constrained by gender. In a study by Male et al. (2018), female engineering students report experiencing a gendered workplace culture that marginalised women and reduced them to feminine stereotypes. Not only female but also male engineering students felt that their agency to act within this gendered workplace culture was limited. In a white-dominated workplace, access and privileges for black people is restricted and impacts on agency (Gyamera & Burke, 2018). Agency is interdependent on power, knowledge and access as well as concepts of gender, age, and race.

The place of agency in future practices is complex and requires individuals to purposefully draw on purpose and choice to see other possibilities, improvise and innovate practices. Using agency at macro, meso and micro levels to consciously amplify purpose and choice of how to act is at the core of the *deliberate professional* (Trede & McEwen, 2016). Deliberate professionals use their agency to enable change. The deliberate professional is someone who is thoughtful yet assertive and action-oriented.

Four Characteristics of the Deliberate Professional

The deliberate professional has four characteristics that each can be understood as a distinct element of agency:

- deliberating on complexity
- understanding what is probable, possible and impossible
- taking a deliberate stance
- being responsible for the consequences of actions.

These four characteristics are interwoven and interdependent, and only together do they comprise the notion of the deliberate professional. Deliberating on complexity is a prerequisite for agency and acting purposefully. It is not possible to purposefully act without an understanding of self, others and the wider context. We concede, of course, that some aspects of practices warrant perpetuations and we have discussed the role of pragmatic agency in the framing agency section of this chapter.

However, to be prepared for future practices takes a pragmatic type of agency that identifies why practices need to change. Complexity implies that practices are not linear and instrumental reasoning alone is insufficient to appreciate that one action can have multiple different reactions depending on how actions are interpreted within contexts and how they are connected to other people and things.

Understanding what is probably going to happen starts from the assumption that norms, codes of conduct and practice traditions shape routine practices and guide professional decision making. It also implies an active awareness of power structures and what will happen when routine practices, especially hegemonic practices that perpetuate injustices, are not interfered with. This characteristic of the deliberate professional is underpinned by critical social practice theory (Bourdieu, 1990). Understanding what is possible taps into agentic creativity that invites individuals to imagine other options and how to practise otherwise. Understanding what is impossible helps individuals to set boundaries for the safety of self and others. Imagination and creativity as well as courage are key capabilities in liquid times to lead future practices. Courage and imagination extend to the consideration of what and how practices will evolve with technology. This characteristic of the deliberate professional is also about curiosity because to understand possibilities starts with asking questions and exploring choices.

Taking a deliberate stance signifies that deliberate professionals don't view practice as being neutral. To be a deliberate professional means using judgement and making purposeful professional decisions. Here agency is the action that affirms what one stands for and strengthens the values that underpin identity agency. Taking a stance demonstrates capacity to think for self and not allow others to make up our minds (Newman, 2006).

Being responsible for the consequences of actions signals that deliberate professionals are deeply reflexive and thoughtful thinkers. Acting in itself is not enough without also considering and observing the effect of actions on self, others and the wider community. Being responsible means understanding the complex context within which actions occur and recognising that solving one problem can open up new problems. Being responsible for the consequences of actions and decisions "nurtures a sense of active citizenship and being an active change maker of unjust, ineffective practices" (Trede & McEwen, 2016, p. 23). Taking actions requires reflecting on their effects. The work is not done by using agency and taking action alone. Being responsible for actions taken involves reflexive agency by learning from these reflections on actions for future actions. This reflexive agency is driven by ethical responsibility to change practices that need improving.

The concept of the deliberate professional adds to an understanding of agency a dimension of autonomous thinking and acting that embraces considerations for self, others and the environment. It positions self as the agent for the social and cultural. Deliberateness for the deliberate professional constitutes a purpose that is underpinned by an interest to uphold and advance ethical, social and sustainable future practices that include but do not privilege economic imperatives. Future practices cannot afford to be driven by one-dimensional interests because they could not be sustained in an increasingly connected, complex and diverse world.

The four characteristics of the deliberate professional are gainfully positioned because they enable professionals to navigate with deliberate agency through liquid times where many ways of how to act are possible. Automated work processes through advances in technology can have negative impacts on professional identity and there can be a loss of agency, self-worth and sense of recognition (Leopold, Ratcheva, & Zahidi, 2018). However, repetitive work processes have never required human agency.

Human capabilities for future practices amplify what it means to be human. These capabilities include critical thinking, imagining other possibilities, emotional intelligence, considering others and being ethically responsible. Deliberate professionals embrace all these human-specific capabilities because they engage purposefully with advances of technology to enhance society and advance democracy. They are mindful that material agency and artificial intelligence can assist but not replace future practices. What is needed is human relational agency that enables optimal use of technological developments to advance humanity and social justice because without these goals in mind the future of our globe will be under threat.

CONCLUSION

Agency can be understood as a capacity of individuals to reproduce, transform or resist social structures and their influences. Seen in isolation, agency would create chaos, placing undue emphasis on individuals. Pursuing agency alone is overly simplistic and ignores context, culture and social structures. Agency and structure are co-dependent and as Kreber (2016) argues, societies need social structures. To an extent we rely on structures and rules for how to live together and remain safe. However, structure alone does not ensure safety and to be human means to have both structure and agency. Deliberately engaging with both these elements is what it takes to move forward into future practices.

Key structures and practices that shape agency include culture, socialisation and social structures. These require the following attributes and behaviours to support and enhance agency: resilience, capabilities and intelligences, and choice. Above all agency requires purpose; without purpose, agency is simply action and potentially chaotic action. Looking towards future practice, deliberate professionals can bring the powerful capacity of agency to shape and overcome future practice challenges.

REFERENCES

Ahern, L. (2001). Language and agency. *Annual Review of Anthropology, 30*, 109-137.

Archer, M. S. (1988). *Culture and agency: The place of culture in social theory*. Cambridge, England: Cambridge University Press.

Bandura, A. (2001). Social cognitive theory: A social agentic perspective. *Annual Review of Psychology, 52*, 1-26.

Bauman, Z. (2012). *Liquid modernity*. Cambridge, England: Polity Press.

Bourdieu, P. (1972/1977). *Outline of a theory of practice* (R. Nice, Trans.). Cambridge, England: Cambridge University Press.

Bourdieu, P. (1990). *The logic of practice*. Cambridge, England: Polity Press.

Bourdieu, P., & Giddens, A. (1984). A theory of structure: Duality, agency, and transformation. *American Journal of Sociology, 98*(1), 1-29

Cornelissen, J. J., & van Wyk, A. S. (2007). Professional socialisation: An influence on professional development and role definition. *South African Journal of Higher Education, 21*(7), 826-841.

Deal, T. E., & Kennedy, A. (1982). *Corporate cultures: The rites and rituals of corporate life*. Harmondsworth, England: Penguin.

De Weerdt, S., Bouwen, R., Corthouts, F., & Martens, H. (2006). Identity transformation as an intercontextual process. *Industry & Higher Education, 20*(5), 317-326.

Dewey, J. (1933). *How we think: A restatement of the relation of reflective thinking to the educative process*. New York, NY: DC Heath and Company.

Fellenz, M. R. (2015). Forming the professional self: *Bildung* and the ontological perspective on professional education and development. *Educational Philosophy and Theory*. doi:10.1080/00131857.2015.1006161

Fuchs, C. (2016). *Critical theory of communication: New readings of Lukács, Adorno, Marcuse, Honneth and Habermas in the age of the internet*. London, England: University of Westminster Press.

George, J. M. (2000). Emotions and leadership: The role of emotional intelligence. *Human Relations 53*(8), 1027-1055.

Giddens, A. (1984). *The constitution of Society: Outline of the theory of structuration*. Cambridge, England: Polity Press.

Gyamera, G. O., & Burke, P. J. (2018). Neoliberalism and curriculum in higher education: A post-colonial analyses. *Teaching in Higher Education, 23*(4), 450-467.

Hall, E. T. (1977). *Beyond culture*. New York, NY: Anchor.

Henschke, K. (2012). The emergence of "I", "you" and "us" identities for, in and through work placements. In S. Debowski (Ed.), *Proceedings of the 35th HERDSA Annual International Conference* (pp. 108-117). Perth, Australia: HERDSA.

Higgs, J. (2013). Professional socialisation including COP. In S. Loftus, T. Gerzina, J. Higgs, M. Smith, & E. Duffy (Eds.), *Educating health professionals: Becoming a university teacher* (pp. 83-92). Rotterdam, The Netherlands: Sense.

Higgs, J. (2016). Deliberately owning my practice model: Realising my professional practice. In F. Trede & C. McEwen (Eds.), *Educating the deliberate professional: Preparing for future practices* (pp. 189-203). Switzerland: Springer.

Higgs, J. (2019). Re-interpreting clinical reasoning: A model of encultured decision making practice capabilities. In J. Higgs, G. Jensen, S. Loftus, & N. Christensen (Eds.), *Clinical reasoning in the health professions* (4th ed., pp. 13-31). Edinburgh, Scotland: Elsevier.

Hitlin, S., & Elder, G. H. (2007). Time, self, and the curiously abstract concept of agency. *Sociological Theory, 25*(2), 170-191.

hooks, b. (1990). *Yearning: race, gender, and cultural politics*. Boston, MA: South End Press.

Knappett, C., & Malafouris, L. (2008). *Material agency: Towards a non-anthropocentric approach*. New York, NY: Springer.

Kreber, C. (2016). *Educating for civic-mindedness: Nurturing authentic professional identities through transformative higher education*. New York, NY: Routledge.

Leopold, T. A., Ratcheva, V., & Zahidi, S. (2018). *The future of jobs report 2018*. Switzerland: Centre for the New Economy and Society, World Economic Forum.

Male, S., Gardner, A., Figueroa, E., & Bennett, D. (2018). Investigation of students' experiences of gendered cultures in engineering workplaces. *European Journal of Engineering Education, 43*(3), 360-377.

Mayer, J. D., Caruso, D. R., & Salovey, P. (1999). Emotional intelligence meets traditional standards for an intelligence. *Intelligence, 27*(4), 267-298.

Mayer, J. D., Caruso, D. R., & Salovey, P. (2016). The ability model of emotional intelligence: Principles and updates. *Emotion Review, 8*(4), 1-11.

Neugenbauer, J., & Evans-Brain, J. (2016). *Employability: Making the most of your career development*. London, England: Sage.

Newman, M. (2006). *Teaching defiance: Stories and strategies for activist educators*. San Francisco, CA: Jossey-Bass.

Reid, A., Dahlgren, L. O., Petocz, P., & Abrandt Dahlgren, M. (2008). Identity and engagement for professional formation. *Studies in Higher Education, 33*(6), 729-742.

Schein, E. (2004). *Organizational culture and leadership* (3rd ed.). San Francisco, CA: Jossey-Bass.

Stephenson, J. (1998). The concept of capability and its importance in higher education, In J. Stephenson & M. Yorke (Eds.), *Capability and quality in higher education* (pp. 1-13). London, England: Kogan Page.

Trede, F., & McEwen, C. (Eds.). (2016). *Educating the deliberate professional: Preparing for future practices*. Switzerland: Springer.

Wacquant, L. (2005) Habitus. In J. Beckert & M. Zafirovski (Eds.), *International encyclopedia of economic sociology* (pp. 317-320). London, England: Routledge.

Youssef, C. M., & Luthans, F. (2007). Positive organizational behavior in the workplace: The impact of hope, optimism and resilience. *Journal of Management, 33*(5), 774-800.

Franziska Trede PhD (ORCID: https://orcid.org/0000-0002-6638-2609)
Institute for Interactive Media and Learning
University of Technology Sydney, Australia

Joy Higgs AM, PhD (ORCID: https://orcid.org/0000-0002-8545-1016)
Emeritus Professor, Charles Sturt University, Australia
Director, Education, Practice and Employability Network, Australia

RUTH BRIDGSTOCK

12. EMPLOYABILITY AND CAREER DEVELOPMENT LEARNING THROUGH SOCIAL MEDIA

Exploring the Potential of LinkedIn

Social media can be a used as a powerful ingredient in career identity development as part of an integrated suite of media tools. LinkedIn, for instance, is known universally as a platform for professional self-promotion and graduate job recruitment. It can also be a valuable tool in 21st century career development learning more broadly, used for research into industry opportunities, structures and norms, professional network development, and informal learning. This chapter explores some of the opportunities for, and uses of, LinkedIn as a platform for specific elements of students' career development and employability learning, and professional network development. It also provides learning examples and discusses pitfalls and contested issues in the use of LinkedIn, and social media more generally, in preparing higher education learners for work.

SOCIAL MEDIA AND LIFE/WORK IN THE 21ST CENTURY

With technology developments (such as smartphones), social media has become a part of everyday life for millions. Social media's original remit related to personal interests, communication and personal relationships, but is now also central to marketing activities, recruitment and work generation/acquisition (Manroop & Richardson, 2013). Social media has transformed the ways in which employers, workers, job seekers and recruiters interact with one another. Adler (2016) surveyed 3,000 workers and found that 85% of them had obtained their most recent roles using social networks. University-aged students (aged 18–24) make up the largest segment of social media users, but a significant number of authors (including the proposer of the term "the digital native" [Prensky, 2011]) now contend that this does not mean that such students are necessarily highly digitally literate, advanced social media users, or use social media for professional purposes. While many universities are now embedding employability skills and career development learning into programs (Bridgstock, 2009; Jackson & Wilton, 2016), there is a need for specific social media learning to be applied to professional and career development (Benson, Morgan, & Filippaios, 2014; Bridgstock, 2019, in press).

LinkedIn was created in 2003 and is now the largest professionally oriented social networking site, with more than 400 million members in 200 countries as of 2015 (LinkedIn, 2015). Like Facebook, Instagram and Twitter, LinkedIn members can create a personal profile for themselves and link with others. LinkedIn profiles

© KONINKLIJKE BRILL NV, LEIDEN, 2019 | DOI: 9789004400795_012 emphasise education and career history, and professional affiliations. Members can connect with other LinkedIn members, thus accessing broader networks. They can communicate directly with these connections. They can join interest and affiliation groups (corporate, conference, networking, industry, professional, alumni and so on), and establish new groups. LinkedIn also enables members to search for employment opportunities, research companies and industries, provide curriculum vitae information, and give and receive recommendations (Buck, 2012). LinkedIn offers members access to a wide variety of articles written by business professionals and other members on diverse topics, but focusing particularly on professional development or on career insights and advice. Nearly all job recruiters use LinkedIn to talent-scout, research and screen potential candidates (Acikgoz & Bergman, 2016). A recent development in recruitment that will become increasingly influential is "talent analytics", which employs big data-based algorithmic approaches to identify suitable candidates on the basis of their published attributes and capabilities, experiences and social networks (Winsborough & Chamorro-Premuzic, 2016).

UNIVERSITY LEARNERS, SOCIAL MEDIA AND LINKEDIN

There is an assumption that because learners in traditional university age groups use social media in their personal lives, they will also do so in their professional lives, and they will be highly adroit at doing so. My survey research with third year undergraduates across two Australian universities (Bridgstock, 2016) revealed that while 82% used Facebook every day, only 12% used LinkedIn regularly. Despite understanding in theory that networking was important to job acquisition, more than eight in ten university graduates surveyed only applied for jobs using direct application methods (e.g. responding to advertisements on SEEK). My research joins that of many others pointing out that a small minority – only 39 million of the more than 400 million members of LinkedIn are students (LinkedIn, 2015), and that social media for professional and career development represents a significant opportunity in university students' career and employability development. Informal use of social media by students does not transfer to confidence or awareness in its applications for work and career (Benson et al., 2014; Pozzi, 2015).

Several studies have investigated why students do not use LinkedIn. Non-users of LinkedIn tend to not see the benefit of joining the network and do not know how to use it or find it confusing (Colbeck, 2015). Dominant in these studies is the finding that students lack skills in using social media for professional purposes. Indeed, skills associated with online technical proficiency skills and professionalism are both lacking in many undergraduate students. Skills for online professionalism include career management, ethics and information security, digital branding, networking and communication online. Manroop and Richardson (2013) found that under-use of LinkedIn for career development can be linked with a "passive and laid back" approach that comprises creation of a profile and then hoping that others (such as employers and recruiters) will seek them out.

McCabe (2017) noted that active participation is a key determinant of success on LinkedIn, including making contact with those with similar interests, active networking and information sharing. Many students who are infrequent or non-users of LinkedIn lack awareness of the features and functions of LinkedIn that can benefit their careers. Manroop and Richardson (2013) interviewed young job seekers, many of whom were completely unaware of the job search features of social media, and some of whom were surprised to learn that social media can be useful for building networks and research. Johnson (2017) further reported that non-members can be surprised to discover that recruiters use social media to review potential candidates. Neier and Zayer (2015) surveyed undergraduate students and found that while many recognised LinkedIn as a tool for building industry connections, most did not perceive it as a medium to gain exposure to others' ideas and opinions, or to facilitate expression of self and ideas.

Another key reason cited for non-engagement or under-engagement of students with LinkedIn is a perception that profile creation is a post-graduation activity, and that students may not have enough work experience to start the process – or even that doing so might damage their careers (Slone & Gaffney, 2016). This finding resonates with other literature suggesting that without scaffolded support, students can leave career development activities until the final semester of study, or even after graduation, disadvantaging their careers and weakening their graduate outcomes. Wetsch (2012) and Bridgstock (2009, 2019 in press) argue strongly for students to commence building a professional network before graduating, since the development of a professional online presence and network takes time. Finally, some students perceive that LinkedIn is not safe, and are concerned about privacy and sharing information about themselves online. While LinkedIn activity tends to be professionally focused and it may therefore be less of a privacy risk than for other social media sites (Zide, Elman, & Shahani-Denning 2014), these concerns may be warranted: social media is linked with a range of privacy and safety issues.

UNIVERSITY TEACHING OF SOCIAL MEDIA FOR CAREER DEVELOPMENT

As social media becomes central to career development, several scholars now argue that professional social media use should be considered a core 21st century capability (Benson et al., 2014; Bridgstock, 2019, in press). While universities are embracing social media to share news and experiment with pedagogy (Erskine et al., 2014; Jenkins et al., 2012), few universities teach social media skills.

Wherever social media is taught, it is often within a specific disciplinary area such as media studies (Kim & Freberg, 2017), where it is examined as a social and technological phenomenon rather than as a set of professional skills. Social media skill development is also taught practically in the context of social media marketing and public relations (e.g. future social media marketers engage in campaign development), and also if social media-based pedagogic strategies are employed for students to learn disciplinary content (Manca & Ranieri, 2016). Social media pedagogy is a growing area of research and practice in higher education, but much less frequently extends to authentic professional activities and career development learning using social media such as LinkedIn (Pozzi, 2015). There are a number of reasons for this: The first is that much of the employability-focused curriculum still emphasises the disciplinary knowledge and skills required for performance "on the job", rather than career development learning. However, this emphasis is changing, with many universities now including career skills in their curricula as well as disciplinary and professional content. The second reason may be related to the fact that some educators lack digital and/or social media skills themselves, or through want of recent experience of industry workplaces, may not be as familiar with current practices and role requirements.

Perhaps the most commonly cited reasons educators give for non-inclusion of social media learning in curricula mirror the concerns of students about risks of privacy, safety and data ownership. The management of these social media risks can be complicated by university social media policies, which may be restrictive, are often complex and multi-faceted, and are subject to change (Pasquini & Evangelopoulos, 2017). Educators also express concerns about a lack of university technical support for their use of social media in teaching, increased academic workloads, needing to separate their own personal and professional online identities, and lack of integration of social media with learning management systems (Manca & Ranieri, 2016). These concerns are valid, and for many represent significant barriers to including social media in their educational practices. However, I join many other commentators in suggesting that the affordances of social media-based learning, and the dominant role that LinkedIn plays for career development in many fields, means that we must find ways to surmount these challenges if we are to provide an effective and up-to-date university learning experience.

LEARNING CAREER AND EMPLOYABILITY CAPABILITIES USING LINKEDIN

In addition to using LinkedIn for career development and employability learning purposes, it can also be used to foster the development of employability capabilities. These capabilities are divided into two categories (refer to Table 12.1): (i) connectedness capabilities focused on the characteristics and affordances of social media and social networks for work and career (Bridgstock & Tippett, 2018), and (ii) career development and employability learning not specific to social media and social networks, but that lends itself to development via the LinkedIn platform, at least in part. The connectedness capabilities presented in Table 12.1 are under-represented in higher education curricula. LinkedIn supports the development of social network literacies and professional branding, and provides stellar opportunities to make professional connections despite this being still seen as primarily the purview of face-to-face interaction.

LinkedIn can also become a site for career identity development and career self-management. Career identity can be thought of as a "cognitive compass", guiding career-related behaviour. It is continually constructed through behaviour, experiences and social interaction. As the expression of skills and career experiences is revised and augmented, and the profile title and headline are adjusted, the student is continually revising their career identity by linking with others with similar professional interests, engaging in active research and interacting with individuals and organisations.

Table 12.1. Categories of learning for career and employability using LinkedIn.

Connectedness capabilities – optimising social networks for career development	
Social network literacy	The ability to reflect upon and articulate (i) the roles that social networks play in professional life, and (ii) how professional social networks operate; the ability to navigate social networks strategically and effectively
Building a connected identity	The ability to represent professional identities effectively in the context of social networks, including social media profiles and personal/professional "branding"
Making connections	The ability to extend and expand professional networks and develop weak ties, including networking
Strengthening and maintaining connections	The ability to strengthen professional connections and develop strong ties, and then maintain these as needed
Working with connections	The ability to work effectively with collaborators in professional contexts and for professional applications, and make use of professional connections for knowledge creation, problem solving, professional learning and career development
Broader career and employability learning through LinkedIn	
Career self-management	The ability to navigate one's career; to define and realise one's personal career objectives based in an adaptive career identity, knowledge of self and the world of work
Professional communication	The ability to share or convey ideas or information accurately, clearly and as intended in professional contexts
Digital literacies	The ability to use digital technology, communication tools or networks to locate, evaluate, use and create information
Digital citizenship	The ability to use digital technology safely, responsibly and professionally, based on knowledge of digital privacy, safety, "netiquette", legal issues and health

Careers educators who use LinkedIn as a career development learning tool often suggest that students develop a title and "I am" statement that links multiple sometimes disparate education and career experiences with an overarching statement about career purpose and activity. Such strategies can make for a more intelligible profile, and mean that students may have to update these profile elements less often; it can also provoke deeper thinking about career identity than typically occurs in the early stages of career.

It is mostly through exposure and experience that members learn how to use LinkedIn strategically for career development. For instance, the combination of LinkedIn with a range of other professionally relevant social media can help with

"Social Klout" and help ensure that students appear in recruitment searches. The choice of social media platforms is industry specific. For instance, in journalism, professionals will often combine LinkedIn with Twitter, and a blog to maximise Klout. Advanced users will also personalise their LinkedIn profiles to reflect their personal "brand", including videos, documents, images and recommendations from others that help to position individual skills and experience (McCabe, 2017).

Other employability skills beyond career development learning that are developed by LinkedIn activity include professional communication and interpersonal skills, intercultural skills, digital literacies, creative capabilities, and privacy and copyright knowledge (McCorkle & McCorkle, 2012). Bridgstock (2019, in press) discusses the wider role of social network capability for the 21st century professional which provides the basis for knowledge creation, and a significant proportion of informal professional learning in knowledge intensive fields. Much of the power of social networks in the 21st century lies in weaker and more indirect ties, crowdsourcing and knowledge networks.

TEACHING APPROACHES AND LEARNING ACTIVITIES

This section suggests learning activities using LinkedIn, grouped by undergraduate degree phases (foundational, broadening and deepening, advanced/capstone). The corresponding career development learning and employability emphases are also described. Ideally, the learning activities should be embedded into a whole-of-program approach, involving increasingly advanced LinkedIn-based activities embedded into a broader employability-orientated curriculum.

Foundational Phase: Exploring Possibilities

A range of issues require early consideration in using social media: the learning emphasis, career development learning goals, privacy, security, legal issues, health, communication, digital literacy, social network literacy, and building a connected identity. In the foundational phases of undergraduate programs, learners may benefit from exploration and investigation of potential career opportunities online, including roles and industry sectors of interest. The outcomes of this exploration can be reflected upon given the learner's values, capabilities and interests, and provisional decisions can be made about careers and plans for study.

Prior to as well as embedded into this exploration should be activities that help ensure that learners will contribute online in ways that are safe, responsible and informed. Pozzi (2016, p. 6) summarises these elements for learners as "protect yourself" (using social media with an understanding of the roles of privacy, security and the law), and "look after yourself" (understanding that the use of social media can affect one's health, and managing this).

As part of a wider research project into career options that may involve interviews with industry practitioners, desk research and brief direct experience of workplaces and work, learners are tasked with researching the online presences of industry-relevant groups, organisations and professionals in their field/s of interest, and sharing their findings with their classmates. They may explore recent organisational and individual activities, jobs posted and the profiles of professionals (in order to identify industry opportunities and trends, required skills and capabilities), and become familiar with professional behavioural norms online. In order to access some of the content required to complete this activity, learners will be required to create their own LinkedIn profiles. The profile creation task can initially be undertaken as a guided classroom activity using a basic "how to develop your LinkedIn profile" guide and existing learner résumés as the basis (McCorkle & McCorkle, 2012). The profile can then be updated progressively as learners develop more advanced knowledge of profile design for their role/s and field/s of interest (Zide et al., 2014). As part of the profile creation exercise, learners should also connect with one another, the program page (if applicable), and the institutional alumni page.

The foundational phase is also an excellent time to develop learners' social network literacies, by using LinkedIn and other social media such as Twitter or Facebook to demonstrate social network phenomena and principles that are useful for professional networking online: such as context collapse; strong, weak and indirect ties; network size and quality; and network complexity and diffusion. Sacks and Graves (2012) presented several useful learning activities for foundational social network literacy development, including an exercise that demonstrates the power of social media by asking learners to use it to map how many degrees of separation they are from a list of influential people, and another that uses analogue methods to map learners' in-class networks to explore the strengths and weaknesses of different network strategies. These foundational social network literacies are useful in the broadening and deepening phase, where learners use social network to strategise their own professional network.

Broadening and Deepening Phase: Connecting with Others

In this broadening and deepening phase, learners focus on honing their LinkedIn profiles, and developing and adding value to their professional networks. An exercise that asks learners to become recruiters shortlisting candidates' LinkedIn profiles for a hypothetical job role may be useful in discovering the ingredients of a highly effective profile (McCorkle & McCorkle, 2012). In this exercise, groups of learners are given 40 LinkedIn profiles and a recent job advertisement. They take approximately 30 minutes to shortlist five candidates for the role. The class then discuss the shortlisting process and outcomes, including the key characteristics of LinkedIn profiles that reflect positively or negatively on the applicant. These characteristics are then embedded into the learners' profiles in the next activity.

A second intermediate activity that draws upon the social network literacy learning from the foundational phase asks learners to draw their existing social networks, and then with their post-university "dream job role" in mind to describe and map how their associated dream professional network would look (Sacks & Graves, 2012). Learners create a plan to move their current network towards their ideal network, including social network characteristics such as complexity, size and quality, and also specific individuals and organisations with whom they wish to connect. Plans include the extent and ways in which the learners plan to use LinkedIn, Twitter and other social media tools such as blogging to meet their networking aims, as well as analogue (face-to-face) approaches.

Once learners have reasonably well-developed LinkedIn profiles and an idea of what they would like their professional networks to look like, they can engage in online and face-to-face networking. For learners who will be engaging in work integrated learning at a capstone level, the networking may be conducted with these opportunities in mind. In this learning activity, learners join LinkedIn groups and connect with organisations and professionals in their chosen career fields. The safety, privacy and legal principles for social media covered in the foundational phase can be revisited at this point, and then extended to ensure that learners are prepared to remain professional when networking. The activity is not just about making professional connections with others; it is also about strengthening connections online and offline through adding value via status updates, sharing content, writing brief articles and contributing to group discussions by asking and responding to questions.

Advanced Phase: Preparing for Launch, Building Identity and Connections

In the advanced phase, learners revise their profiles again, with a view to the next phase of their journeys (in many instances this will be post-graduation initial job search, but not always). By this stage, periodic and ongoing revision and updating of social media profiles should be accepted as normal. An effective learning activity for advanced learners is a SWOT (Strengths, Weaknesses, Opportunities, Threats) analysis (Johnson, 2017). Thinking of their next career steps and with professional differentiation in mind, learners reflect upon their current professional strengths and weaknesses in terms of skills and capabilities, the market demand/trends for their preferred job roles and industries, and their major external threats (including competitors). They provide specific evidence for their answers, including examples/experiences from work and learning, and industry trend research from relevant and reliable sources. Learners can then use their SWOT analysis to summarise their most outstanding accomplishments, and add this to the summary section of their LinkedIn profile (McCorkle & McCorkle, 2012).

Using their knowledge of successful social media strategies undertaken by others in their professional field/s, learners can also create their own portfolio of social media profiles in addition to LinkedIn. Each platform can add in different ways to learners' professional narratives (Johnson, 2017). For instance, professionals with a visual communication-related role may benefit from using Instagram, Pinterest or YouTube. Learners who use platforms beyond LinkedIn should learn the principles and practices of social media reputation management. Sites such as BrandYourself and AboutMe can be useful in increasing search engine rankings for professional social media profiles, and can be used for tracking and analytics.

CONCLUSION

Learning to use social media for professional purposes is an ongoing process that requires a commitment to continual exploration, connection with others, and revision of identity and brand. It is also here to stay. LinkedIn is now a key means by which people acquire work, and connect with others for professional learning and career development. LinkedIn can also be a useful platform for career development and employability learning more broadly. Thus, it should be included in higher education programs, and the recognised concerns and pitfalls around its use should be considered and addressed. Many of these can be ameliorated through explicit learning of digital literacies and digital citizenship not only by learners and teachers, but also by educational leadership and policymakers.

REFERENCES

Acikgoz, Y., & Bergman, S. M. (2016). Social media and employee recruitment: Chasing the runaway bandwagon. In Landers R & G. Schmidt (Eds.), *Social media in employee selection and recruitment* (pp. 175-195). Cham, Switzerland: Springer.

Adler, L. (2016). *New survey reveals 85% of all jobs are filled via networking*. Retrieved from https://www.LinkedIn.com/pulse/new-survey-reveals-85-all-jobs-filled-via-networking-lou-adler/

Benson, V., Morgan, S., & Filippaios, F. (2014). Social career management: Social media and employability skills gap. *Computers in Human Behavior, 30*, 519-525.

Bridgstock, R. (2009). The graduate attributes we've overlooked: Enhancing graduate employability through career management skills. *Higher Education Research & Development, 28*(1), 27-39.

Bridgstock, R. (2016). *Why take a networked approach to graduate employability? For students and graduates*. Retrieved from www.graduateemployability2-0.com

Bridgstock, R. (2019, in press). Graduate employability 2.0: Learning for life and work in a socially networked world. In J. Higgs, G. Crisp, & W. Letts, (Eds.), *Education for employability I: The employability agenda*. Rotterdam, The Netherlands: Brill Sense.

Bridgstock, R., & Tippett, N. (2018). *Higher education and the future of graduate employability: A connectedness learning approach*. London, England: Edward Elgar.

Buck, S. (2012). *The beginner's guide to LinkedIn*. Retrieved from http://mashable.com/2012/05/23/LinkedIn-beginners/

Colbeck, K. (2015). *Navigating the social landscape: An exploration of social networking site usage among emerging adults* (Unpublished master's thesis). University of Western Ontario, Canada. Retrieved from http://ir.lib.uwo.ca/etd/2815

Erskine, M., Fustos, M., McDaniel, A., & Watkins, D. (2014). *Social media in higher education: Exploring content guidelines and policy using a Grounded Theory approach*. Paper presented at the Twentieth Americas Conference on Information Systems, Savannah, GA.

Jackson, D., & Wilton, N. (2016). Developing career management competencies among undergraduates and the role of work-integrated learning. *Teaching in Higher Education*, 21(3), 266-286.

Jenkins, G., Lyons, K., Bridgstock, R. S., & Carr, L. (2012). Like our page: Using Facebook to support first year students in their transition to higher education. *International Journal of the First Year in Higher Education, 3*(2), 65-72.

Johnson, K. M. (2017). The importance of personal branding in social media: Educating students to create and manage their personal brand. *International Journal of Education and Social Science*, 4(1), 31-27.

Kim, C., & Freberg, K. (2017). The state of social media curriculum: Exploring professional expectations of pedagogy and practices to equip the next generation of professionals. *Journal of Public Relations Education*, 2(2).

LinkedIn. (2015). *400 million members: Why LinkedIn is so important for our students and future graduates.* Retrieved from https://socialmediaforlearning.com/2015/11/03/400-million-members-why-LinkedIn-is-so-important-for-our-students-and-future-graduates/

Manca, S., & Ranieri, M. (2016). Facebook and the others: Potentials and obstacles of social media for teaching in higher education. *Computers & Education, 95*, 216-230.

Manroop, L., & Richardson, J. (2013). Using social media for job search: Evidence from Generation Y job seekers. In T. Bondarouk & M. R. Olvas-Lujan (Eds.), *Social media in human resources management* (Vol. 12, pp. 167-180). Bingley, England: Emerald Group Publishing Limited.

McCabe, M. B. (2017). Social media marketing strategies for career advancement: An analysis of LinkedIn. *Journal of Business and Behavioral Sciences, 29*(1), 85.

McCorkle, D. E., & McCorkle, Y. L. (2012). Using LinkedIn in the marketing classroom: Exploratory insights and recommendations for teaching social media/networking. *Marketing Education Review, 22*(2), 157-166.

Neier, S., & Zayer, L. T. (2015). Students' perceptions and experiences of social media in higher education. *Journal of Marketing Education, 37*(3), 133-143.

Pasquini, L. A., & Evangelopoulos, N. (2017). Sociotechnical stewardship in higher education: A field study of social media policy documents. *Journal of Computing in Higher Education, 29*(2), 218-239.

Pozzi, M. (2015). 'Create a better online you': Designing online learning resources to develop undergraduate social media skills. *International Journal of Social Media and Interactive Learning Environments, 3*(4), 305-321.

Pozzi, M. (2016, February). *Formalising the vernacular: Social media skills in higher education.* Paper presented at VALA 2016: Libraries, Technology and the Future, Melbourne, Australia.

Prensky, M. (2011). Digital wisdom and homo sapiens digital. In M. Thomas (Ed.), *Deconstructing digital natives: Young people, technology, and the new literacies* (pp. 15-29). New York, NY: Taylor & Francis.

Sacks, M. A., & Graves, N. (2012). How many 'friends' do you need? Teaching students how to network using social media. *Business Communication Quarterly, 75*(1), 80-88.

Slone, A. R., & Gaffney, A. L. (2016). Assessing students' use of LinkedIn in a business and professional communication course. *Communication Teacher, 30*(4), 206-214.

Wetsch, L. R. (2012). A personal branding assignment using social media. *Journal of Advertising Education, 16*(1), 30-36.

Winsborough, D., & Chamorro-Premuzic, T. (2016). Talent identification in the digital world: New talent signals and the future of HR assessment. *People and Strategy, 39*(2), 28-31.

Zide, J., Elman, B., & Shahani-Denning, C. (2014). LinkedIn and recruitment: How profiles differ across occupations. *Employee Relations, 36*(5), 583-604.

Ruth Bridgstock PhD (ORCID: https://orcid.org/0000-0003-0072-2815)
Centre for Learning Futures
Griffith University, Australia

JOY HIGGS AND DANIEL RADOVICH

13. RE-IMAGINING PRACTICE STRUCTURES AND PATHWAYS

Starting to Realise Tomorrow's Practices Today

> Creativity is a central source of meaning in our lives … most of the things that are interesting, important, and human are the results of creativity. … when we are involved in it, we feel that we are living more fully than during the rest of life. (Csikszentmihalyi, 2013, pp. 1-2)

Creativity and vision are the starting point for this chapter. People in future-oriented jobs and companies are moving beyond traditional practices to make work more future-relevant, future-viable and future-creative. Traditional social and economic structures shape organisations and the way many people work or practise. This chapter challenges the predictability, restrictions and future suitability of these "givens". It emerged from reflections on various questions:

- What if communities have needs that current services or jobs do not meet?
- What if society and clients are missing out on services that non-traditional workers or entrepreneurs can offer?
- What if the affordances of new technologies and artificial intelligence (AI) are underutilised in today's workforce?
- What if the rate of change in an industry requires high levels of job and worker flexibility; is this definitely a negative state of affairs?
- How might we re-think issues of job security without reverting to old answers like unemployment benefits and under-utilisation of talent and enthusiasm?
- What are the barriers to future-oriented changes in practice and workforces and how might these be overcome?
- How might we imagine industries and workforce systems to carry work and workers forward into evolving practice futures?

Taking these questions as a launching pad, this chapter will explore a number of issues and trends in the re-imagining of practice. These will be illustrated with examples (in italics) from the work-life experiences of one of our authors (Daniel).

NEW WORK SYSTEMS

If we look back over the last 20–30 years (at the most), what jobs were rare (new) or didn't yet exist? In this section a range of new systems of working and organising work are presented. While labelled to reflect their key characteristics, these systems frequently overlap in organisational approach, jobs and clientele.

Knowing about these new systems could promote understanding of new work patterns by previous generations, it could prompt new workers to look beyond established work arenas and modes, and it could provide stepping stones to even newer approaches.

The Gig Economy and the Sharing Economy

Many new labels and approaches are appearing in today's shifting cultural and business environment. The gig economy, for instance, can be thought of as a free-market system in which temporary positions are common and organisations, employers and/or clients contract short-term engagements with independent workers. This economy matches providers to consumers on a gig/job basis in support of on-demand commerce. De Stefano (2016) describes two main forms of work in the gig economy. The first is "crowdwork", which involves completing tasks through online platforms. The second is "work-on-demand via app" in which apps (applications) are used to connect workers to traditional work like transport, cleaning, clerical work, etc.

Kalleberg and Dunn (2016) discuss the rapid growth of new economies which include: the sharing economy, collaborating economy, crowdworking, access economy, on-demand economy, freelance economy, 1099 economy and platform economy, among other terms. These authors alert us to the positives of such jobs, such as promotion of entrepreneurship and innovation, jobs with flexibility, autonomy and work-life balance, and opportunities for individuals to monetise their resources (e.g. their time, talents, minds, physical abilities, cars, computers). Kalleberg and Dunn also identify potential problems such as worker disenfranchisement, worker exploitation and lack of a social safety net (including health insurance, portable retirement benefits and wage insurance).

The sharing economy focuses on collaboration. This could be across consumers and workers and indicates the ability and perhaps preference of individuals to rent or borrow goods rather than buy and own them. Cusumano (2015) examines the way the sharing economy challenges and competes with traditional companies. He sees the new companies as Web platforms that bring together individuals with underutilised assets with people who would wish to rent those assets short-term. Examples of these assets include spare time for everyday tasks (e.g. TaskRabbit, Fiverr), spare time and automobiles to drive people around (e.g. Uber, Lyft), spare rooms (Airbnb, Flipkey, Roomba) and such items as used tools and household items (Streetbank, Snap-Goods/Simplist). According to Cusumano (2015), sharing-economy startups can be seen as a logical outgrowth of social media platforms such as Facebook, Pinterest and Trip Advisor, which bring together people who have common interests to share information, ideas or personal observations. These startups threaten traditional companies through the capacity of the new companies with their peer-to-peer networks to grow exponentially.

Entrepreneurial Jobs

Entrepreneurship is the creation of new enterprises (Low & MacMillan, 1988). It implies growth, innovation and flexibility; it can apply to both people and organisations (Stevenson & Jarillo, 2007). Successful entrepreneurship has the goal and outcome of creating value (Sharma & Chrisman, 2007). While entrepreneurship, under various names, has been around for centuries, in the 21st century it appears to be increasingly more common and potentially in the future it will be the norm or a necessary part of all work and organisations.

e-Companies

The 21st century has brought a major change in global economic and business conditions that involves a realignment from traditional corporate structures to Internet-leveraged, brand-owning, customer-focused companies. Means and Schneider (2000) describe this as "metacapitalism". Such companies, he argues, are accelerating economic growth and value creation by capitalising on:

- global expansion of market access
- improvements in the efficiency of capital markets
- better leverage of capital
- dramatic unleashing of human potential and capital
- significant advances in operating efficiency.

"Brand-owning" companies, as opposed to manufacturers, develop controls and systems to guarantee that their network partners are well integrated with each other and the marketplace. And the "value-added communities" support brand owners in dramatically reducing costs, increasing quality and responding rapidly to customer demand and market shifts. The e-business industry has boomed in the 21st century with many successful start-ups being formed in North America, Europe, China and India, in particular (Garbade, 2007). Examples include Google, Yahoo, MSN, YouTube, MySpace, Facebook, Photobucket, Studiverzeichnis, Skype, Xing, Bebo, Last.fm, Gumtree, Kijiji, Joost, Tradera, Alibaba, Taobao, Xiaonei, Tom Online, etc. These businesses are characterised by mergers, acquisitions and incorporations.

Artificial Intelligence (AI) Jobs

Nilsson (1998) presents a progression of AI systems or "agents": from elementary AI agents that react to sensed properties in their environments, to evolutionary AI agents that can use techniques beyond sensing to exploit information about the task environment. AI "is concerned with intelligent behavior in artifacts. Intelligent behavior, in turn, involves perception, reasoning, learning, communicating, and acting in complex environments. AI has as one of its long-term goals the development of machines that can do these things as well as humans can or possibly even better" (ibid, p. 1). AI may be a potential threat, an inevitability or a highly beneficial part of the workforce of the future. Stephen Hawking, for instance, while acknowledging the value of AI to date, warned that AI that can match or surpass humans could end mankind (Cellan-Jones, 2014).

Advances in Artificial Intelligence technology and related fields have opened up new markets and new opportunities for progress in critical areas such as health, education, energy, economic inclusion, social welfare, and the environment. In recent years, machines have surpassed humans in the performance of certain tasks related to intelligence, such as aspects of image recognition. Experts forecast that rapid progress in the field of specialized artificial intelligence will continue. Although it is unlikely that machines will exhibit broadly-applicable intelligence comparable to or exceeding that of humans in the next 20 years, it is to be expected that machines will continue to reach and exceed human performance on more and more tasks. (Executive Office of the President of the United States, 2016, n.p.)

Consequences and Implications

There are numerous consequences, implications and changes needed to many aspects of life and society arising from these new modes of work. New privacy and copyright laws have been required to deal with online communications, retirement benefits have been significantly affected, the cost of unemployment to the public and private purse is high, tertiary education has required constant reinvention and so on.

Of particular concern in the labour market is the growth of vulnerable work, which accounts for 1.5 billion people or over 46% of total employment (ILO, 2017). Vulnerable workers are likely to experience volatile income, limited access to social protections, poor working conditions, little career development and greater exposure to unethical behaviours including harassment and bullying … The conditions which lead to vulnerable work include fierce competition for work and networked forms of employment, both of which are common in sought-after graduate occupations as graduate numbers rise and the number of traditional full-time positions declines. (Bennett, 2019, in press)

NARRATIVE 1: CREATING CONTEMPORARY CAREERS AND COMPANIES

Where do contemporary careers (journeys of work and professional development rather than traditional "one-job" careers) and companies start? How are they created by people with vision to see beyond the status quo and imagine future practice?

Daniel: New Industries – Being Part of the Adventure

My name is Daniel Rad and I'm a Sydney based creative. Here are some extracts from my website.[1] ***My goal is to make peoples' lives easier through beautiful design.*** *I've always been a passionately curious person. It gives me a lot of joy spending my days working on interesting business problems. Since landing my first creative role at 16, I've become ever curious about the world of design. It is when* *working as a digital designer at Charles Sturt University that I began to form a practical appreciation of human-centred design. Leading a site build for a research branch of the university at 19 years old, compelled me to consider aspects of design that reached beyond that of visual aesthetics.*

I completed a bachelor's degree in Communications and Media with a major in marketing, communications and advertising. My goal was to get a job working with technology with an emphasis on design and creativity. I also wanted to work in a situation where I had a degree of autonomy relevant to my level of experience, and opportunities to make decisions that shaped the final product of my work. I see myself as a worker who is agile and constantly adaptive not fixed in one job area.

My early jobs were amateur positions that utilised design software. These jobs included working as a photo re-toucher and graphic designer, a digital designer working on print products and a web designer. From these beginnings I entered the world of entrepreneurial companies. I chose to seek work in companies that I thought would be willing to respond to the style of work and diverse skills set that I have. In particular, I approached and considered companies like the following: fast growing start-ups and companies that build disruptive products (in terms of software, systems of consultancy). Such disruption takes the form of development of value-altering products and services in established fields that are 10 times better than the established services (e.g. they have shorter set up timelines, are more portable and have better outcomes). Hence they disrupt the market and create something (a product or a service) that is better than existing options. These companies also challenge the taken-for-granted decisions and level of satisfaction of established companies and they awaken users to new and more successful products. Their competitors sometimes change, compete and flourish, while others continue to pursue existing practices/products and consequently may fail to remain competitive.

Here's a quote I like:

> Driven by opportunism (why stay at a company where advancement opportunities are limited?) and necessity (what else can you do when your job is outsourced?), the practice of switching jobs and companies grew more common, until job-hopping became the norm. (Bersin, 2017, p. 67)

In the early days I did quite a bit of job hopping in order to jump up in pay quite rapidly, develop a very broad range of skills and knowledge and gain good relationships and networks. I was able to do this without damaging relationships with previous companies. This was due to two reasons: first, in the eight months (or so) that I worked in each job I gained the respect of my employers for the work I had contributed, and second, in this field of communications technology, this job mobility is very common and mutually beneficial. Workers gain an increasing range of skills and knowledge (like I have done) and the employers are able to tap into a changing array of talent and capital in their workers that brings repeated injections of new ideas, design capacity and products to the company's human/resource capital. This situation is very different to established work arenas where products and projects are typically longer-term endeavours and where building people's capacity to perform "the way we do it here" is a priority. In my field it is easy to say that knowledge and technology (in the information and communication technologies [ICT]) changes quickly. But this is much more than just a saying: our technologies advance exponentially, they don't just have a short shelf/use life, they have a short development life, with new products and services constantly appearing on the market; and they can re-shape whole companies rather than being one of many products. Further, the workers in these companies are especially fast-moving. In three months, for instance, a team in this industry can create new services, contribute to improvements in and publicly release market-leading products, and pop-up companies can be born.

Reflecting on my early work, I realise how much I learned in these settings and how much I realised that I needed to keep learning. I'll talk more about this below. Also I realised how much flexibility and mobility was part of this world. Employers both expected workers to move on and were pleased to welcome new talents. And I recognised how much each new job opened up learning opportunities and demands. I rapidly gained new ideas and skills and I developed greater work capital and employability. I moved between jobs to explore different work options and expand my work capital and employability.

Observation. Consider the language of the new generation in Daniel's words. He is presenting to his audience as a person, not just a worker or service provider, someone to engage and network with, as well as someone open to new work partnerships. On both counts his words resonate with the spirit of insightful, genuine engagement.

Three new labels to describe careers relate to the story Daniel is presenting. These are the "protean career" (Hall, 2004; Hall & Moss, 1998), the "boundaryless career" (Eby, Butts, & Lockwood, 2003) and "dispositional employability" (Fugate, 2006; Fugate, Kinicki, & Ashforth, 2004). Each of these career types/ways of working involve people: utilising self-efficacy, having the ability to remain flexible and adaptive, demonstrating the capacity to make appropriate work-related choices within a constantly changing employment landscape and directing their own career relatively independently of employers' requirements or government needs.

SUCCESSFUL WORK ARENAS – ORGANISATIONS, WORKPLACES

What does it take to have a successful organisation in the 21st century? Lombardi (2018) says that the global business environment (and we would argue many other industry areas) in the 21st century demonstrates a higher degree of ambiguity than in the past; it is characterised by market uncertainty, digitalisation, information sharing, the end of lifetime careers, commoditisation of services and products, and open talent economies. She contends that organisations wishing to prosper in this context must bring fresh strategy approaches, resilience in the face of ever-changing marketplaces and unprecedented creativity, and they need to work to retain both talent and customers. She refers to the work of James Murphy, founder of Engage International[2] and The Lotus Awards[3] who nominated four interactive pillars in the "Colours

Model" he developed around effective companies: culture, innovation, sustainability and customer experience. Blending these colours (or pillars) produces "white light", i.e. companies that are better aligned and more productive, and their employees have a greater sense of purpose, says Murphy. These pillars are underpinned by trust and a learning culture. Lombardi (ibid) identified six common components of great companies that provide a framework for successful 21st century organisations: vision, values, practices, narrative/story of the organisation, people and place. Another view of successful 21st century organisations is provided by Yeramyan (2014), who identifies six pillars of such organisations: relentless innovation, being purposeful, passion for growth, customer oneness (thinking like, and working in partnership with, clients), having a 21st century leader (see below) and creating breakthrough environments (that "flex to support the rapid movement, speedy decision-making and alignment required to outperform competitors and regularly achieve extraordinary outcomes" (p. 3). Challenging times!

NARRATIVE 2: WHAT'S CHANGED? HOW IS IT WORKING?

How do people grow their careers and companies from their initial entry phase? What are the sorts of challenges they face and motivations that drive them forward?

What am I doing these days? Extracts from Daniel's website

Currently working as a hybrid UX[4] (user experience designer), UI (User interface designer) & Front End Developer[5] (working on software technology specific to things that are visible to the technology user), my hands are in numerous aspects of the spectrum that make up User Experience Design. I spend my days designing flows, wireframing and crafting crisp user interfaces. In my spare time I love playing guitar in indie bands, reading books and speaking at meet-ups.
My goals and philosophy are epitomised in the following roles.

Human-Centred Design: *It is the user at the other end of the product who matters most. My process of design strives to produce refreshing digital experiences that excite emotions and satisfy needs.*

Information Architecture: *Good IA allows a user to seamlessly navigate through a digital experience finding exactly what they need. Good IA factors in heuristics and the desire of the user to create a familiar and frictionless experience. When forming IA I often use primary research such as interviews and surveys or secondary research such as reading support tickets and studying web analytics. My current company offers niche products and operations to our clients in the area of Influence Marketing. Influencers are like micro celebrities who have a highly visible online presence. Most of the time their followings occur as a result of creating interesting content on their social media channels. Influencers try out the product or service and agree via a brief/contract to promote the product to their followers, in their own way.*

> *Influencers have spent much energy and time building up their followings and our business harnesses this networking and sales potential. So we are placing the product via social media, through the influencers, in contact with potential users/purchasers. This system also allows us to track and assess the product placement itself, the product attraction to the potential audience and the measure of audience engagement. And, certainly this is a much more meaningful connection with the potential market than a static billboard, for instance, sitting beside the road.*

Observation. Three key things that Daniel is gaining through his employment and education pursuits are employ-ability (particularly enhanced role diversity, flexibility and capability) plus confidence and future employment vision.

INDIVIDUAL EMPLOYABILITY – WHAT SORT OF WORKERS ARE NEEDED

> Graduates will need to know how to create their own work, pitch ideas to clients and manage their time and development. They will experience multiple ways of knowing and doing, multiple cultural contexts and multiple environments. They will frequently be hired on a by-project or labour hire basis. … work requires them to be knowledge workers whose practice is enabled by a capacity and willingness to learn. Graduates will need to know how to work autonomously, possibly from home, possible alone, for at least part of their working lives. Graduates will need to be prepared for inconsistent work which will require them to define and redefine 'success' in terms of their ability to create and sustain meaningful work and a living wage. At least initially, they are unlikely to have much say about when and where flexibility occurs. The negotiation of these schemata is employability development in the truest sense. (Bennett, 2019, in press)

In the changing and innovative spaces of practice, work and new organisational structures, the notion and realisation of employability is a central matter. Older ideas of meeting employer recruitment checklists, matching organisational or industry stereotypes and bringing predictable skills and qualifications to the job are no longer the key issue. Re-thinking employability in the contexts opened up in this chapter is valuable. Consider, for this following idea from Fugate et al. (2004):

> We examine the idea that an individual's employability subsumes a host of person-centered constructs needed to deal effectively with the career-related changes occurring in today's economy. We argue that employability represents a form of work specific (pro)active adaptability that consists of three dimensions—career identity, personal adaptability, and social and human capital. (p. 14)

Career identity deals with how people define themselves in their work context, make sense of their past and present work, and frame direction for their future career. It provides the cognitive-affective foundation for meaning making.

Personal adaptability involves changing personal parameters (such as optimism, propensity to learn, openness and self-efficacy) to meet the demands of the situation. Adaptability benefits both the worker and the organisation or company. An internal locus of control is central to personal adaptability.

Social and human capital relates to the goodwill and capacities inherent in social networks and the capacity of individuals to identify and realise career opportunities. People can invest in their own social/human capital through developing new capabilities, pursuing education and learning on-the-job, and across jobs. Companies can invest in their workforce social and human capital through recruitment of capable personnel and staff development.

> career identity, personal adaptability, and social and human capital exert a mutual influence on each other. None of these factors operate independently; to understand the implications of any given factor and of employability as a gestalt, one must examine the entire constellation of factors (cf. Bandura, 1977). (Fugate et al., 2004, p. 27)

Bersin (2017) speaks of 21st century graduates as often having skills not found in "experienced hires"; these graduates are frequently asked to manage mature workers and are expected to "re-skill" themselves. He compares traditional careers that were characterised by expertise/profession/identity, duration and rewards with today's jobs where expertise has an ever-shorter shelf life, single long-lasting careers are a "thing of the past" and rewards are indefinite. Bersin sees success in work today as a practice of "surfing from wave to wave"; this involves entering a good job (catching a wave) and riding it to the crest then moving on to the next wave before the first one falls. Managing the challenges and benefits of wave surfing, argues Bersin, is both a management and worker responsibility. Higher paid jobs, he argues, are going to people with high levels of highly (currently!) marketable technical skill areas and to those performing hybrid (or renaissance) jobs that "create whole new job categories by mashing up disciplines ... [and] combine technical expertise (in one or more domains) with expertise in design, project management, or client and customer interaction" (ibid, p. 72). Examples include "experience architects", "user experience designers" and "security consultants".

Looking at the links between learning and work, a pattern emerges. On-the-job learning is of ancient origins. Traditional education–work links persist, including formal education, followed by work, followed by more formal education (e.g. a graduate development degree for specialisation or extension, or a postgraduate degree for depth, or a second entry degree for change of profession). Self-directed learning has ever been the pursuit of the motivated. However, three new learning for work options are reflected in the terms *micro credentialing* (gaining informal recognition of study/learning certificates to extend knowledge and abilities beyond formal curricula), *e-portfolios* (records of informal learning often accompanied by reflections on that learning, supplementary to formal degree testamurs) and *learning alongside work*. For instance, Mirvis and Hall (1996) developed the term *learning a living* to refer to periods of work interspersed with periods of learning, or concurrent work and learning. Such learning allows the flexible worker to more preparedly move between jobs, and to use jobs for learning.

LEADERSHIP IN THE 21ST CENTURY

According to Yeramyan (2014), a 21st century leader is one who is authentic, open and brings out the best performance, expression and creativity in everyone. Such leaders create breakthrough environments that "flex to support the rapid movement, speedy decision-making and alignment required to outperform competitors and regularly achieve extraordinary outcomes" (p. 3). Once leaders create leader–team alignment, "they can successfully navigate ambiguity at all levels and pursue the biggest possibilities for the organization" (p. 3).

For Kozłowski and Bratnicki (2015), passion is the key and driving force for entrepreneurial success and it is a powerful force for accomplishing goals. They see entrepreneurial leadership as follows:

> Forming vision and creating strategic direction, transforming organizational culture, and mobilizing motivation constitute the basic elements of entrepreneurial leadership. (p. 16)

Daniel: Pursuing my Vision

Two things that I think have played a big part in my work journey are learning and autonomy. I think that the idea of "learning a living" (see Mirvis & Hall, 1996, above) suits me well. This approach to work requires me to be flexible and to recognise how much my working world and my own abilities are rapidly changing. It is a particular point of conversation among the highly technical community that change in our field occurs at a very high rate. This means that constant learning is vital for my work enjoyment, progression and success. While I see my long-term future caught up in IT I don't see myself as pursuing a single career; ICT is the platform not the fixed description upon which my work is evolving. And, I know that there will continue to be a high demand for technical people which gives me confidence in my job security and stability. It's a field of work, not one job.

When I think about the type of worker this field of work needs, I can see two broadly different types of workers, and neither has a fixed set of technical skills. If they did then this would be wasteful on the one hand, for instance, it might take 3–5 years to build up a high level of technical expertise in one specific programming language. However, well before that time had elapsed the industry could have moved on to new programming language making the previous one obsolete. First there are those who do jobs like project management; they need to have expertise in soft skills like being able to collaborate with a lot of different stakeholders, inspire their teams and manage time. Their abilities require an understanding of business, sales and productivity; they need granular skills like time management and attention to detail. Their hard (technical) skills are limited, but they need to know what their team members and colleagues in different sections of the company can deliver. Second, there are the more technical roles like back end developers whose expertise is very technical; they predominantly have hard skills like writing complex code. They can "get results from a computer" and have excellent logic to the point of how a physicist would formulate an argument. In either case prospective employers of both types of worker would want to see evidence of the quality work they had produced in the past, going far beyond any qualifications.

All the people who are advancing in my field learn through their work and are constantly learning. We have to dedicate serious time to learning both inside and outside work hours, spending a minimum of 1–2 hours per day perfecting our craft and knowledge and improving our ability to contribute creatively to our work. We engage in online learning throughout our careers to avoid becoming obsolete. Without this we won't have expertise or be up-to-date. Plus, we're continually branching out and learning new technologies.

For myself, apart from my degree and learning through different jobs – at one stage three at once – I have completed three private courses of 10 weeks with classes twice a week. These were User Experience Design (conducted by Academy XI), Front End Web Development (conducted by General Assembly) and Java Script Development (conducted by General Assembly). Each brought me valuable learning, new skills and insights.

What do I see as my emerging leadership abilities and potential? In my field the organisational structure is flatter rather than hierarchical. This means that leadership is more about what we do and the way we do it, rather than a step up the ladder. It's about being in the position to lead (through assigned tasks and personal capabilities) rather than having a particular senior title. Leadership skills matter if you want to be in a position where you're collaborating with a lot of people and educating others and helping others collaborate well. And, leadership skills are needed if you want to be more highly paid.

For myself I have a lead design role. That came about through two things: being thrown in a deep end, in being tasked with developing products or solving problems with a team of people; and because I made it clear that I was keen to be challenged and lead in ventures that are complex and in unknown areas. Being willing to lead and take responsibility, having knowledge and intuition have set me up well to lead. In a way I have stumbled into an industry that has these high demands, I have had a high degree of luck with the people who have offered me these challenges, I have been pulled into such roles and I've been willing and successful in these tasks. My industry sometimes finds it difficult to get people who are willing to take on these roles and challenges. So it's appreciated when people volunteer to lead.

It's interesting too that I enjoy and seek out roles that give me a large degree of autonomy. Of course this doesn't mean I can do what I want regardless of the company's purpose and interests. But it does mean – and I appreciate this greatly – that I am able to have a lot of freedom in creativity in my tasks, that being in an emerging field where I need to create ideas and solutions because there is no set pattern or pathway and also because while science and technology underpin the capacity to do this work, there is a strong acceptance that creativity is building the products in the new future of our work.

CONCLUSION

We return now to the questions at the start of this chapter and offer these conclusions.

- People participating in work or practice – both practitioners and clients as well as other stakeholders – bring their own interests, interpretive lenses and frames of reference to decision making and actions in shaping practice futures. These are illustrated in Figure 13.1.

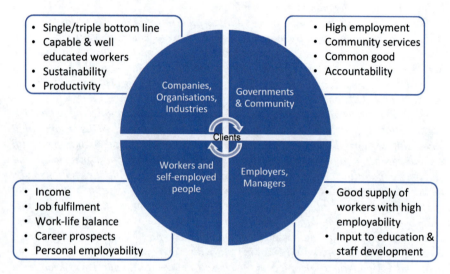

Figure 13.1. *Groups, agencies: Their interests and sources of change.*

- Drivers for change (e.g. dissatisfaction with the status quo, lack of services to particular groups or areas of need, changes in technology, new partnership opportunities) may stimulate the introduction of new work and product possibilities, new structures and pathways for change. Motivated people very often "find a way" through political, commercial, professional and innovative action to make things happen that they feel passionate about. For instance, practitioners may approach non-typical budgeters or charity groups for funding to introduce client services (such as therapy in schools) that the usual first line of action agencies (e.g. government departments) would block through policy decisions or lack of funding for the whole client population.
- Businesses that are entrepreneurial, creative, innovative and more risk tolerant than their traditional counterparts are often acting one or many steps ahead of others. They may well be creating products and services that clients don't yet know that they want, but often wonder afterwards – however did I do without this for all these years?
- A powerful driver for positive change (but also sometimes for negative change and change for change's sake) is technological and knowledge advances which

provide opportunities and affordances for whole new ways of thinking about and doing work. Dynamic technology-savvy workers and thought leaders are the ones who often capitalise on these opportunities while others are yet to learn how to "read the manual".
- People's lives can be radically affected by changes – that were/are unimaginable in our recent yesterdays and imminent tomorrows. Whole new coping strategies are needed for those made vulnerable by work-demanding flexibilities and unpredictabilities that impact on income reliability, changes in professional identity, etc. Who is helping workers and the yet-to-be employed to prepare for life–work changes as well as work–life changes?
- Clients and the community are not – and don't have to be – simply recipients of change. Becoming involved in practice futures and planning of practice futures is vital for future life and services available. Seeds can be planted by needs, ideas and requests from clients as well as ideas and sales pitches from service providers and entrepreneurs.
- Knowing more about AI provides avenues for using and sharing its advantages rather than worrying about its job takeover possibilities.
- Time between jobs may bring unemployment and devastation or learning and self-redirection time.

We have reflected in this chapter on new jobs, new work structures and practice strategies. This is a fascinating space for observation, inspiration and participation.

NOTES

[1] https://www.danielrad.com.au/
[2] http://www.engageinter.com/
[3] https://www.lotus-awards.com/
[4] UX is a big term and it can refer to a user experience: consultant, researcher or practitioner/generalist.
[5] Compared to back end jobs which relate to operations, security and stability.

REFERENCES

Bennett, D. (2019, in press). Meeting society's expectations of graduates: Education for the public good. In J. Higgs, G. Crisp, & W. Letts (Eds.), *Education for employability I: The employability agenda*. Rotterdam, The Netherlands: Brill Sense.

Bersin, J. (2017). Catch the wave: The 21st-century career. *Deloitte Review, 21*, 62-79.

Cellan-Jones, R. (2014, December 2). Stephen Hawking warns artificial intelligence could end mankind. *BBC News*. Retrieved from https://www.bbc.com/news/technology-30290540

Csikszentmihalyi, M. (2013). *Creativity: The psychology of discovery and invention*. New York, NY: Harperennial Modern Classics.

Cusumano, M. A. (2015). How traditional firms must compete in the sharing economy. *Communications of the ACM, 58*(1), 32-34.

De Stefano, V. (2016). *The rise of the 'just-in-time workforce': On-demand work, crowdwork, and labor protection in the 'gig-economy'*. Geneva, Switzerland: International Labour Office.

Eby, L., Butts, M., & Lockwood, A. (2003). Predictors of success in the era of the boundaryless career. *Journal of Organizational Behavior, 24*(6), 689-708.

Executive Office of the President of the United States. (2016). *Artificial intelligence, automation, and the economy*. Washington, DC: US Government.

Fugate, M. (2006). Employability. In J. Greenhaus & G. Callanan (Eds.), *Encyclopedia of career development* (pp. 267-270). Thousand Oaks, CA: Sage Publications.

Fugate, M., Kinicki, A., & Ashforth, B. E. (2004). Employability: A psycho-social construct, its dimensions, and applications. *Journal of Vocational Behavior, 65*(1), 14-38.

Garbade, M. J. (2007). *International mergers & acquisitions, cooperations and networks in the e-business industry*. Retrieved from https://papers.ssrn.com/sol3/papers.cfm?abstract_id=1291375

Hall, D. (2004). The protean career: A quarter-century journey. *Journal of Vocational Behavior, 65*(1), 1-13.

Hall, D., & Moss, J. (1998). The new protean career contract: Helping organisations and employees adapt. *Organizational Dynamics, Winter*, 22-37.

International Labour Organization (ILO). (2017). *World employment and social outlook: Trends 2017*. Geneva, Switzerland: International Labour Office.

Kalleberg, A. L., & Dunn, M. (2016). Good jobs, bad jobs in the gig economy. *Perspectives on Work*, 10-13, 74.

Kozłowski, R., & Bratnicki, M. (2015). New insights into entrepreneurial leadership concept inspired by passion. *Journal of Global Academic Institute Business and Economics, 1*(2) 7-21.

Lombardi, G. (2018, January 19). The four defining pillars of the successful 21st century organisation. *Marginalia: Future of Work Magazine*. Retrieved from http://www.marginalia.online/the-four-defining-pillars-of-the-successful-21st-century-organisation/

Low, M. B., & MacMillan, I. C. (1988). Entrepreneurship: Past research and future challenges. *Journal of Management, 14*(2), 139-161.

Means, D., & Schneider, G. (2000). *Metacapitalism: The e-business revolution and the design of 21st-century companies and markets*. New York, NY: John Wiley.

Mirvis, P. H., & Hall, D. T. (1996). New organizational forms and the new career. In D. Hall (Ed.), *The career is dead – long live the career* (pp. 72-100). San Francisco, CA: Jossey-Bass.

Nilsson, N. J. (1998). *Artificial intelligence: A new synthesis*. San Francisco, CA: Morgan Kaufmann.

Sharma, P., & Chrisman, S. J. J. (2007). Toward a reconciliation of the definitional issues in the field of corporate entrepreneurship. In Á. Cuervo., D. Ribeiro, & S. Roig (Eds), *Entrepreneurship* (pp. 83-103). Berlin, Germany: Springer.

Stevenson, H. H., & Jarillo, J. C. (2007). A paradigm of entrepreneurship: Entrepreneurial management. In Á. Cuervo., D. Ribeiro, & S. Roig (Eds), *Entrepreneurship* (pp. 155-170). Berlin, Germany: Springer.

Yeramyan, P. (2014, September 3). The six defining traits of the successful 21st century organization. *Forbes*. Retrieved from https://www.forbes.com/sites/gapinternational/2014/09/03/the-six-defining-traits-of-the-successful-21st-century-organization/

Joy Higgs AM, PhD (ORCID: https://orcid.org/0000-0002-8545-1016)
Emeritus Professor, Charles Sturt University, Australia
Director, Education, Practice and Employability Network, Australia

Daniel Radovich, Bachelor of Comms & Media
UX, UI & Front End Developer

NOEL MALONEY

14. FREELANCING, ENTREPRENEURSHIP AND INHERENT CAREER RISK

An Exploration in the Creative Industries

Understanding the practice of particular professions and industries is not just about looking at how to break this practice down into the talents and capabilities that are needed to do the job. We need to take a big picture view of an industry, understand its distinguishing characteristics and individual career pathways, and in so doing, better educate graduates to enter these occupations and to shape their future careers.

This chapter is built on two key observations. Some careers inherently exist and operate through pathways of risk and uncertainty; entrepreneurship and freelancing are common practice while job security is rare. This topic is the focus of my chapter. In other cases some jobs and careers today have little or no current history; they are emerging in the current era of de-stabilisation of career predictability and their commonality lies in the creativity of the entrepreneurs and risk takers who are building these new jobs. Whether they become careers or fleeting occupations is another question, which I leave to other writers.

This chapter portrays the central role that risk, along with freelancing and entrepreneurship, plays in the practices and pathways of the creative arts industries. Risk is a lingua franca for professional artists. Concepts of risk permeate the way they experience, imagine and undertake the work they do. Although risk management is widely encountered across most industries and professions (Lupton, 2013), artists encounter risk in ways that set them apart from other professions. They risk failure, and at times their reputation, to produce work of meaning and gain recognition but at the same time, they need to mitigate and manage the many economic and personal contingencies that challenge them. For artists, these two forms of risk, creative and economic, are related: to research and experiment in an art form, an artist needs the time, place, people and resources to do so in a supportive environment (Balshaw, 2014; Dempster, 2014).

In structuring this chapter, I first consider two occupations in the creative industries: screenwriters and actors. Next, I examine the nature of freelancing work and entrepreneurship. Finally, I reflect on risk in these work contexts: risk as *a philosophical reality, that is part of this world*, risk in relation to career management and risk as a personal challenge that may or may not be comfortably met in the lived reality of individual creative artists.

THE WORLD OF SCREENWRITING

Screenwriters work either on contracts for production companies as script readers, script editors or story developers, or as freelancers. In many ways, the specific practice of screenwriting is emblematic of how many artists work in general. Like other artists, they generate an artefact for sale, in their case a treatment, scene breakdown or script. Screenwriters typically work across a range of production platforms, including television serials, series, telemovies and films. Screenwriters will often be working concurrently on several assignments, including contracted or commissioned scripts, as well as their own original projects. The writers interviewed in research referred to in this chapter typically juggled a range of writing contracts and projects, and in some cases, also taught writing.

Screenwriting differs from many other art work forms by being a highly industrialised form of creative labour (Bloor, 2012; Maras, 2009). Screenwriting work on television serials and series is fragmented, with clear divisions between script writing, script editing and script producing. In feature film writing, early career writers may lack the status to see a production through from beginning to finish (Conor, 2014). One of the biggest risks facing freelance screenwriters is the extent to which television and film markets are fickle, and subject to change. Feature film projects are difficult to finance, and developmental funding for scripts competitive.

Australian cinema historically attracts low audiences, and for the past 10 years has only averaged 4.1% of the total domestic box office (Screen Australia, 2017). For the last five years, there have been on average of 40 Australian television drama programs, or 449 hours, produced annually (Screen Australia, 2017). In 2016/2017 there was a resurgence in drama series production, but other than *Neighbours* and *Home and Away*, there were no long-form productions. Overall, television drama series in Australia are now shorter in length and below five hours in total (Screen Australia, 2018). Typically, these projects favour more experienced writers who can be trusted to deliver. As a consequence, entry-level opportunities for newer writers are less available than they once were.[1] In addition, there are only six television broadcasters in Australia. It is no surprise then, that Australian screenwriters are seeking work in the United Kingdom, Europe and the United States. As the writers interviewed for my research on screenwriters' work note, one of the chief motivators for seeking employment outside of Australia was to work in regions that offered more opportunity, and more choice. As Anna,[2] one of the writers interviewed, put it, when you pitch a television project to Australian networks, after six "no's", there are no more opportunities.

It is argued that, historically, screenwriters have low creative autonomy (Maras, 2016). Writers contracted to work on existing television productions will be required to develop scene breakdowns and script drafts according to a prescribed narrative style. At the same time, they must demonstrate impressive storytelling skills through pitching and story development in script and production meetings. In this regard, a screenwriter's acting skills are as important as their writing skills (Caldwell, 2008). As MacDonald (2013) argues, film and television scripts are not simply "authored texts". Instead, screenwriters participate in a complex process that is guided by a set of time-honoured production rituals. Writers are expected to be able to pitch a screen narrative to potential investors or buyers within a few short, sharp sentences that capture plot, genre and audience appeal.

When working on a television series, writers will participate in collectivised story development or plotting sessions. This aggregated form of screenwriting labour enables production companies to maintain narrative and budgetary control, and provides a range of efficiencies (Caldwell, 2008). As Caldwell notes, collective development sessions promote creativity, but they also produce competition, "one-upmanship", long work hours and high levels of anxiety. For screenwriters, the inherent industrialisation of their largely independent, creative, pre-production or inception-oriented work means that they experience their work as contradictory. It is both individualised and collaborative; it is rewarding but competitive and offers a rich mixture of both pleasures and pain (Conor, 2010).

Bloor argues that most film making is restricted to the generic expectations of the market, and risky, "transformational" or so-called artistic work is discouraged because of the large investments at stake (Bloor, 2012). By contrast, there is a strong tradition of craft in screenwriting. Richard Sennett (2006) defines craft as doing something well for its own sake. Its cardinal value is commitment and for Sennett, getting something right, even though it will give you nothing, is the spirit of true craftsmanship. There is a certain rationality implied in Sennett's description of craft. To do something well for its own sake requires deliberation and focus. Such an approach also requires planning.

Caldwell (2008) characterises the expression of craft principles by practitioners as "industrial self-theorizing" and argues that it provides a logic for daily writing practice and serves the needs of screen production industries. He goes on to argue that while craft principles have theoretical origins, their constant, habitual use has rendered them natural and normative. Certainly, there is a strong suggestion in the interviews that the craft language used by these writers is part and parcel of the work they do on a daily basis. Drawing on interviews she did with UK screenwriters, Conor argues that the industrial emphasis on narrative structure provides a form of discipline that is, to as certain extent, welcomed by screenwriters. While rigorous, it is also "fun, comforting and pure of form". It provides a sense of certainty, especially when writers are redrafting scripts (Conor, 2014). However, it is also argued that the craft philosophy of slow, deliberate and ongoing work enables exploitation. Concepts of "quality" and "craft", along with a philosophical framing of creative labour as "lifelong learning", have been critiqued as a way of disguising exploitation and promoting the fragmentation of labour (McRobbie, 2016).

THE CREATIVE ARTS AND RISK

To examine the way creativity and risk can be and need to be balanced in the creative industries I interviewed 10 screenwriters with at least five years of professional screenwriting experience (Maloney, 2019, in press).[3] A number of strategies were identified that these writers used to pursue this balance:

Being Entrepreneurs

The 10 interviewees demonstrated a sophisticated awareness of patterns in film and television production economies. They promoted themselves in three specific ways: they develop screen projects for sale or as a promotional tool; they maintain and extend professional networks; and they consciously build reputations.

Calling Cards

Each of the interviewees had one or more projects in development, either in partnership with other writers or producers, or as sole copyright ventures. Aside from developing this work for sale, they valued polished projects as useful "calling cards", with which they could promote their writing style, and secure future work.

Social Networking

Although they had a low opinion of their networking skills overall, the interviewees did identify specific forms of social networking they considered useful. These included networking at industry events and attending training workshops where they were more likely to make meaningful contacts with other writers and producers.

Reputation Building

Reputation building is a core entrepreneurial activity that can occur in two keys ways. Screenwriters build a reputation for reliability by delivering powerful stories, on time, within production requirements. As well, they accumulate screenwriting credits. However, interviewees noted that reputation building is a fragile activity, requiring time and care.

Intuition

As well as consciously applying these entrepreneurial principles to career management, the interviewees valued intuition in their decision making. Sensing when to make a career move, feeling a connection with one's writing and reading the emotional climate in production companies were considered important.

Keeping the Wheel Turning

Approaches to work and time management varied. A business-like approach to one's writing was considered important, but so too was a capacity to manage one's emotional health. All of the writers interviewed managed multiple assignments, and time management was crucial. For some, a well-managed whiteboard was essential. For others, an ability to feel one's way with workloads was the preferred approach.

Relationship Building

Only one interviewee voluntarily mentioned collaboration as a component of their creative practice. Most writers focused on the need to build productive relationships, with effective listening, building trust, imagining the needs of others and providing mentorship as key components of this practice.

Craft Ethos

There is a strong tradition of craft within screenwriting. As Conor (2014) argued, it is a way of screenwriters distancing themselves from more refined notions of the artistic or the literary. The interviewees supported this way of conceiving their work. As well, the language of craft enables these writers to develop complex aesthetic judgements about their own work, and that of others.

THE MANY DIFFERENT PATHS OF ACTING

Unlike screenwriting, acting has no trouble conceiving of itself as an art. The actor as artist is a concept central to acting training (Chekhov, 2013; Hagen & Haskel, 1973; Stanislavski, 2003). With it comes the notion that actors must take risks and remain vulnerable in order to better their art (Seton, 2004).

However, the economic risks for actors are not so well taught. In a study of 782 Australian actors, more than 75% of respondents said their acting training provided no instruction on financial management (Maxwell, Seton, & Szabó, 2015). Yet, the employment figures for actors in Australia are stark. In their study of Australian professional artists, Throsby and Petetskaya (2017) estimate that there are around 7,900 actor and directors in Australia in 2016.[4] Their survey makes a standard distinction between core, creative work; arts-related work such as teaching; and non-related work. In the financial year 2014/2015, actors and directors averaged $19,600 gross p.a. from core creative work, $12,800 p.a. from arts-related work, and $13,700 p.a. from non-related work. In total, their average income was $46,100 p.a. This compares to the average Australian wage in 2014/2015 of $61,600 p.a. While a small number of actors and directors earn high wages, 51% earned less than $10,00 p.a. from core, creative work (ibid).

There is a disparity between the time actors and directors spend on their core, creative work, and the income they earn from it. Each week, Australian actors and directors averaged 19 hours on their core creative work (not all of which is paid), 15 hours on arts-related work and 7 hours on paid, non-arts related work (ibid). These artists are spending most of their time on their creative work, but earning less from it than they are from non-arts related work. There will be several reasons for this. Actors' wages are low and not all of this work will be paid. In addition, work hours may include time actors spend privately rehearsing for a performance, attending classes, and performing in a profit-share theatre production or film that generates little net income for participants. As Melbourne-based performer and director Maude Davey notes, "profit share" often means "loss share" (cited in Watts, 2015).

Full-time acting positions are scarce. Fremantle Media's *Neighbours* and Channel 7's *Home and Away* are the only ongoing Australia television drama productions. Actors may secure parts in long-running musicals or be fortunate enough to join a funded theatre ensemble, although permanent companies of waged actors are almost an economic impossibility in Australia (Milne, 2008). Overall, the vast majority of professional Australian actors freelance, with jobs in one-off theatre productions with subsidised or commercial theatre companies, and short-term roles in television serials or series, feature films, advertisements, or corporate and educational videos.

Given the freelance nature of theatre performance, theatrical agents are commonly seen as an important service for performers. They can secure auditions, assist with contracts, represent their clients' interests and overall, provide valuable information on work opportunities. However, the Media, Entertainment and Arts Alliance (MEAA), the union that covers performers, is lobbying for tighter regulation of performer agents to prevent exploitation of clients (MEAA, 2016).

Australia's extensive independent theatre sector also provides opportunities for actors and directors to develop work. The sector comprises a large and highly diverse range of performance companies such as MKA, Lab Kelpie, Little Ones, Version 1.0 and Bakers' Dozen, to name just a few. These companies may receive occasional project development grants but are generally not fully funded. As a result, their work may be intermittent. These companies will have a small core of creative directors and will hire actors on a project basis. Profits, if any, will typically be shared. Importantly, the independent sector is supported by specific venues such as La Mama Theatre in Melbourne, PACT and the Griffin Theatre in Sydney and La Boite in Brisbane, as well as a range of development programs such as the HATCH accelerator program offered through the Frankston Arts Centre in Melbourne.

15 Minutes from Anywhere, a Melbourne-based artistic collaboration between director Beng Oh and writer Jane Miller, is an independent theatre company that has been actively producing work in Melbourne for the past eight years, as well as touring productions nationally. Unlike many other independent companies, it maintains several productions in repertoire, including Miller's play, *Cuckoo*, which commences a third production at North Melbourne's Meat Market in August 2018. The company's roll call of associated artists is indicative of the highly collaborative nature of independent theatre, with many of those listed also actively involved with other companies in Melbourne (15 Minutes from Anywhere, 2018).

As a testament to the importance of independent theatre and the challenges it poses in terms of remuneration, MEAA has recently developed a new independent theatre agreement that legally acknowledges actors as volunteers who, while not formally paid, may still receive an honorarium (Watts, 2017). As MEAA organiser Erica Lovell notes, such an agreement is important because independent theatre provides actors with an opportunity to sharpen their skills, to network and to promote themselves (cited in Watts, 2017).

This summary of the performing arts sector in Australia, though brief, suggests that there are specific self and career-management skills actors require in order to work well in it. The sector comprises complex networks of freelance actors constantly moving from one project to another. The ability to efficiently navigate these networks, form effective bonds, cultivate relationships and maintain reputations are essential abilities. The balancing of creative, arts related and non-arts related work that the majority of professional Australian actors undertake implies a different range of self-management skills from other occupations. Time management is an obvious skill needed but actors need a particular type of non-arts related work that allows them the flexibility to take on performance commitments when they occur. Acting work is typically inconsistent, and at times unpredictable. An actor might be cast in a production at short notice; this will require them to temporarily forgo other work.

Conversely, actors might also spend lengthy periods without acting work. Given this unpredictability, how do actors nurture their creative identities, especially as they grow older? In a study of 16 actors in their 40s and 50s, Felix Nobis observed that the unpredictable nature of acting work challenged actors' professional identities in a range of ways (Nobis, 2015). Many of those he interviewed expressed a deep satisfaction with their work, but for some, the absence of acting work meant a lack of completeness.

Others were more accepting of realities as they grew older. Notably, some participants in this study were entrepreneurially applying their acting skills in other areas, such as corporate training. Nonetheless, those who continued to think of themselves as actors but sought financial security in training or teaching could still become uncertain about their acting identities (ibid). This dilemma, between the desire to perform and the need to earn a viable living, is a common experience for professional artists in general, with 53% claiming insufficient income as the main problem preventing them increasing time spent on their core, creative work (Throsby & Petetskaya, 2017).

ENTREPRENEURSHIP

Entrepreneurship and social networking are now considered to be of central importance for professional artists growing and sustaining their careers (Bridgstock, 2014). Once thought to be the result of innate character traits, entrepreneurship is now seen as involving particular ways of thinking (Duening, 2010) that include assessment, judgements and decision making to evaluate opportunity, and create and grow ventures (Mitchell et al., 2002). Building on Howard Gardner's essential mind model for effective human health (Gardner, 2007), Duening (2010) proposed a model encompassing five "minds" for successful entrepreneurial futures:

- the Opportunity Recognising Mind recognises useful patterns in human behaviour, economies and resources
- the Designing Mind brings a new service or product to the market, or discerns an existing opportunity

- the Risk Managing Mind manages and minimises risk effectively
- the Resilient Mind thrives despite turbulence, change or trauma
- the Effectuating Mind creates something of value and delivers it to a market.

Bridgstock (2014) reminds us that arts professionals will shape entrepreneurial behaviour and thinking in ways that are distinct from business more generally, and will be influenced by specific contexts. Where entrepreneurship has been taught in creative arts programs, it has typically been delivered in a "business school approach" that often lacks relevance to arts contexts (Bridgstock, 2012, p. 132). However, creative art practice and entrepreneurship can co-exist and be mutually beneficial (Beckman, 2007). Bridgstock (2012) recommends an incorporation of entrepreneurship education within creative arts programs, that is staged progressively and allows students to adapt career identities through an iterative and reflective process, in conjunction with the development of disciplinary and technical skills and knowledge. Such an approach, she argues, should include opportunities to create ventures or projects that are congruent with personal values, research case studies within students' fields, and eventually, to co-create individual projects. Such an approach builds on the concept of "doing entrepreneurship" (Raffo et al., 2000). The strength of Bridgstock's model lies in the progressive development of career identify, coupled with a staged, "safe" exposure to real-world contingencies. It emphasises project-based learning and collaboration, and proposes these pedagogies be integrated with discipline-based skills.

Such an approach would certainly be beneficial to actor training, in that it would enable acting students to better research real-world experiences, as well as develop flexible creative identities with which they can better negotiate the economic challenges their profession presents. Indeed, the MEAA study into actors' health and wellbeing referenced earlier specifically recommends that training institutions prioritise career planning (Maxwell et al., 2015).

Bridgstock's suggested approach also matches the experience shared by the screenwriters I interviewed. Their creative practice goes hand in hand with the need to be entrepreneurial and to manage their work effectively. Nonetheless, as the screenwriters have clearly demonstrated, the successful management of a creative practice is not an entirely rational endeavour, with relationships, intuition and aesthetic judgement playing crucial roles in effective decisions and collaboration.

CONCLUSION

This chapter portrays ways in which screenwriters and actors manage their creative practices. One of the defining characteristics of both professions is the level of freelancing and entrepreneurship their work demands. Each profession also requires a level of creative risk be taken, in order for its practitioners to produce work of value. At the same time, these professions entail significant economic risks, and challenge practitioners to plan their work effectively. They require the concrete skills of time and project management, and ability to identify opportunities, as well as the "softer" skills of relationship building and communication.

These professions are also subject to a range of influences beyond the control of their practitioners. Television and film markets are volatile, while theatre, television and film production are highly dependent on government funding, which has contracted nationally in Australia. Some people in the creative arts industries, such as the screenwriters referenced in this chapter, have responded to these challenges by developing global careers, but not all have such opportunities. Certainly, a flexible approach not only to employment, but also to one's creative identity, is essential. How then will creative arts education and training respond to these challenges?

NOTES

[1] Fremantle Media's long-running serial drama *Neighbours* has just announced that it will be reinstating an in-house story team, who will be responsible for plotting episodes and story arcs. This is good news for an industry in desperate need of training opportunities for new writers.
[2] Pseudonyms are assigned to the research participants for anonymity.
[3] The interviews, part of a study funded by the Victorian VET Development Centre to research the skills and knowledge needed by professional Australian screenwriters to develop international careers, were undertaken in 2015 and 2016.
[4] These figures are drawn from *Making Art Work: An Economic Study of Professional Artists in Australia*, which was funded by the Australia Council for the Arts, undertaken by Macquarie University, and based on the Australian census data gathered in 2016.

REFERENCES

15 Minutes from Anywhere. (2018). *About*. Retrieved from http://15minutesfromanywhere.com/about/

Balshaw, M. (2014, June). *Maria Balshaw on cultures of risk management*. Presented at The Art of Risk symposium, University of Leeds, England.

Beckman, G. (2007). 'Adventuring' arts entrepreneurship curricula in higher education: An examination of present efforts, obstacles, and best practices. *Journal of Arts Management, Law, and Society, 37*(2), 87-112.

Bloor, P. (2012). *The screenplay business: Managing creativity and script development in the film industry*. Abingdon, England: Routledge.

Bridgstock, R. (2012). Not a dirty word: Arts entrepreneurship and higher education. *Arts & Humanities in Higher Education, 12*(2-3), 122-137.

Bridgstock, R. (2014). Professional capabilities for twenty-first century creative careers: Lessons from outstandingly successful Australian artists and designers. *The International Journal of Art and Design Education, 32*(2), 176-189.

Caldwell, J. T. (2008). *Industrial reflexivity and critical practice in film and television*. Durham, NC: Duke University Press.

Chekhov, M. (2013). *To the actor on the technique of acting*. London, England: Taylor & Francis.

Conor, B. (2010). 'Everybody's a writer': Theorizing screenwriting as creative labour. *Journal of Screenwriting, 1*(1), 27-43.

Conor, B. (2014). *Screenwriting: Creative labor and professional practice*. Abingdon, England: Routledge.

Dempster, A. (2014). *Risk and uncertainty in the art world*. London, England: Bloomsbury.

Duening, T. (2010). Five minds for the entrepreneurial future: Cognitive skills as the intellectual foundation for next generation entrepreneurship curricula. *The Journal of Entrepreneurship, 19*(1), 1-22.

Gardner, H. (2007). *Five minds for the future*. Cambridge, MA: Harvard Business School Press.

Hagen, U., & Haskel, F. (1973). *Respect for acting*. New York, NY: Macmillan.
Lupton, D. (2013). *Risk*. Oxford, England: Routledge.
MacDonald, I. (2013). *Screenwriting poetics and the screen idea*. New York, NY: Palgrave Macmillan.
Maloney, N. (2019, in press). Understanding employability in the creative industries. In J. Higgs, G. Crisp, & W. Letts (Eds.), *Education for employability II: Learning for future possibilities*. Rotterdam, The Netherlands: Brill Sense.
Maras, S. (2009). *Screenwriting: History, theory and practice*. London, England: Wallflower Press.
Maras, S. (2016). *Ethics in screenwriting: New perspectives*. London, England: Palgrave Macmillan.
Maxwell, I., Seton, M., & Szabó, M. (2015). The Australian actors' wellbeing study: A preliminary report. *About Performance, 13*, 69-113, 233-235.
McRobbie, A. (2016). *Be creative: Making a living in the new cultural industries*. Cambridge, England: Polity Press.
Media, Entertainment and Arts Alliance (MEAA). (2016). *Opening address to Victorian inquiry into the labour hire industry and insecure work*. Retrieved from https://www.meaa.org/mediaroom/opening-address-to-victorian-inquiry-into-the-labour-hire-industry-and-insecure-work/
Milne, G. (2008). Lighthouse: A 'mainstage' ensemble experience. *Australasian Drama Studies, 53*, 42-57.
Mitchell, R. K., Busenitz, L., Lant, T., McDougall, P. P., Morse, E. A. & Smith, J. B. (2002). Toward a theory of entrepreneurial cognition: Rethinking the people side of entrepreneurship research. *Entrepreneurship Theory & Practice, 27*(2), 93-104.
Nobis, F. (2015). Dropping a part: The changing relationship of midlife actors with their profession. *About Performance, 13*, 197-210, 233.
Raffo, C., Lovatt, A., Banks, M., & O'Connor, J. (2000). Teaching and learning entrepreneurship for micro and small businesses. *Education + Training, 42*(6), 356-365.
Screen Australia. (2017). *Fact finders: Cinema*. Retrieved from https://www.screenaustralia.gov.au/fact-finders/cinema/industry-trends/box-office/australian-box-office
Screen Australia. (2018). *Screen Australia drama report: Production of feature films, TV and online drama in Australia in 2016/17*. Retrieved from https://www.screenaustralia.gov.au/getmedia/38aef7ec-ed7d-4423-85e6-c247fb6a2066/DramaReport-2016-2017.pdf
Sennett, R. (2006). *The culture of the new capitalism*. New Haven, CT: Yale University Press.
Seton, M. (2004). *Forming (in)vulnerable bodies: Intercorporeal experiences in sites of actor training in Australia* (Unpublished doctoral dissertation). University of Sydney, Sydney, Australia.
Stanislavski, C. (2003). *An actor prepares*. New York, NY: Routledge.
Throsby, D., & Petetskaya, K. (2017). *Making art work: An economic study of professional artists in Australia*. Retrieved from http://www.australiacouncil.gov.au/workspace/uploads/files/making-art-work-throsby-report-5a05106d0bb69.pdf
Watts, R. (2015). *The pros and cons of profit share*. Melbourne, Australia: La Trobe University.
Watts, R. (2017). *Equity develop new agreement for indie theatre*. Melbourne, Australia: La Trobe University.

Noel Maloney PhD (ORCID: https://orcid.org/0000-0001-8862-6435)
School of Humanities and Social Sciences
La Trobe University, Australia

STEVEN CORK AND JENNIFER MALBON

15. YOUNG PEOPLE'S HOPES AND FEARS FOR THE FUTURE

WHY CONSIDER YOUNG PEOPLE?

The vast majority of literature on possible futures documents the thoughts of older, often male, "experts", who are unlikely to experience the future they speculate about. This is as true of literature on futures of work and workers as it is of other futures-thinking literature. It is true that many of these older people have great experience and insights, based on years of learning and experience, both within disciplines and in the "classroom of life". But it is also true that the attitudes, including hopes and fears, of young people about to enter tomorrow's workforces will increasingly shape the future, especially if the future that they inherit does not fit comfortably with their expectations. Future possibilities for occupational practice, in particular, will be influenced strongly by how young people perceive the attractiveness of occupations not only in terms of their careers and financial security but also their ethical and moral values.

Possible futures, in general, can both excite and scare young people. For some, the process of educating oneself to enter the workforce is a daunting prospect, and many truncate that process and find whatever ways they can to support themselves and survive (Malbon, 2017). At the other extreme, many young people are driven to choose an occupation as a career and find they have to deal with very competitive and stressful environments to succeed (Foley et al., 2016).

Richard Eckersley, in his research on "fear of the Apocalypse" in young people, concluded that there are three responses to concerns about the future: *nihilism* (live for today); *fundamentalism* (find a person or a movement that provides simple guidelines for dealing with uncertainty); and *activism* (take action to make the world more like what you would like it to be) (Eckersley, 2008). Nihilism is likely to be unhelpful if workplaces require long-term commitment from workers and/or education and training that require such commitment. Some forms of fundamentalism might be consistent with future work demands but others may not, and fundamentalist (i.e. grossly simplified) thinking is, in itself, likely to be unhelpful when dealing with complex and demanding challenges. Activism often carries with it enthusiasm, creativity, independence and other qualities that might be required in future workplaces, but it could also be a problem if workplaces have not considered the type of world that those entering the workforce expect or desire.

Chapter 3 of this book considered trends that might shape the future of work and practice. This chapter explores what has been documented about young people's attitudes towards the future of work and the challenges that current and future generations might face in the workforce, particularly in relation to work that might require ongoing learning and development (for practice as defined below).

Our starting point is the first phase of the *Making Our Future Work* project (Malbon, 2017) (see below). Although the project is at an early stage, it has already raised several important questions relevant to this book, such as "what factors are likely to encourage or discourage young people when thinking about their working life?" This chapter, however, goes beyond this starting point to review other surveys from around the world about young people's hopes and fears about the future. We consider how these views might influence whether or not young people choose occupations that require long-term knowledge and skills development.

DEFINING YOUNG PEOPLE AND PRACTICE

We define *young people* as those about to enter the workforce or in the early stages of their working life. In most cases, this equates to being 16–31 years of age at the time their views were sought. The studies that we review deal with people born between the 1920s and 2002. McCrindle and Wolfinger (2009) identify six Australian generations during this period: the Federation Generation (born 1901–1924), Builders (1925–1946), Baby Boomers (1947–1964), Generation X (1965–1979), Generation Y (also called Millennials) (1980–1994) and Generation Z (1995–2009). We consider *practice* to mean *occupational practice*, which:

> … encompasses the various practices that comprise occupations, be they professions, disciplines, vocations or occupations. For doctors, engineers, historians, priests, physicists, musicians, carpenters and many other occupational groups, practice refers to the activities, models, norms, language, discourse, ways of knowing and thinking, technical capacities, knowledge, identities, philosophies and other sociocultural practices that collectively comprise their particular occupation. (Higgs, 2012, p. 3)

We take *practice* to imply a significant commitment to education, training and long-term development of skills and experience, and we explore factors that might encourage or discourage this degree of commitment in a range of plausible futures.

THE *MAKING OUR FUTURE WORK* PROJECT

We were inspired to write this chapter by our involvement in the *Making Our Future Work* project (Malbon, 2017), hosted by the independent think tank Australia21.[1] This project seeks to gain an understanding of young people's experiences and feelings about the world of work, their future, and the challenges and opportunities they face. A key assumption is that understanding the attitudes of young people preparing to enter the workforce, or in the early stages of working for a living, is important both in terms of helping those young people deal with the, sometimes considerable, uncertainty surrounding work and careers and also in terms of designing workplaces that will attract and retain a workforce.

The pilot stage of the project was a workshop in September 2017, in the Australian Capital Territory, designed in consultation with an expert reference group and facilitated by young people (aged 31 or under). Fourteen people aged 16–24 years, from a range of backgrounds (employed, unemployed, students, unstable housing), participated in the 3.5-hour workshop. Some were on a path towards possible occupations and some were taking whatever work was available. Activities included an individual survey, small group discussion and whole group discussion. The workshop's focal questions were not designed to elicit opinions about occupational practice in particular, but about work in general.

Four key topics emerged from the dialogue: education; social equity; workers' rights; and career paths. Across all topics, participants talked about the considerable change they perceived to be occurring in society and their feelings of uncertainty about what the future might be like and their roles in it. The hope most frequently cited was to find rewarding, stable and secure work. The most cited fear was overwhelmingly about instability and insecurity of employment. Most participants expressed uncertainty about what opportunities for employment might be available in the future and how different employment pathways might support desired lifestyles. Even those who were pursuing ongoing career development towards occupations (i.e. through university study) made comments suggesting that the pathways towards their desired occupation were unclear and uncertain. Of particular significance to this book was the observation by several workshop participants that they had made decisions about their education and training before they realised what the long-term implications might be. This observation is very concerning, given that we view practice as investment in longer-term education and training for occupational reasons. Although the sample size in this study is small so far, the emerging conclusions are consistent with a range of other studies in Australia and other countries, which are reviewed below.

GENERAL ASPIRATIONS AND ATTITUDES OF YOUNG PEOPLE AROUND THE WORLD

Numerous surveys of the attitudes of young people have been published around the world since the 1970s. Especially since the late 1990s and early 2000s, these surveys have focused on the generation about to enter, or just having entered, the workforce. Before summarising these studies, we note the regular conclusion by social researchers that young people's views about the future are often poorly developed and tend to reflect the dominant narratives that they hear from media, older community members and their peers (Hicks, 1996; Loughlin & Barling, 2001). In addition, it has been noted regularly since the 1970s that young people from primary and secondary school age to late adolescence have shown a tendency to see personal futures separately from wider-world futures. Views about personal futures typically have reflected an expectation that they will have lives like recent generations, whereas views about global futures have followed pessimistic outlooks circulating within society (Hicks, 1996). For example, surveys in the 1980s and early 1990s, in the UK, USA and Australia, found that teenagers' concerns for the world reflected widely publicised concerns of the times, including poverty, inequality, unemployment, environmental disasters and major conflicts, but expectations for personal futures included secure employment, marriage, establishing a family and being better off and happier than their parents (Brown, 1984; Henley Centre, 1991; Hicks, 1996; Hutchinson, 1992; Johnson, 1987).

To an extent, most recent surveys are consistent with the generalisations made above. Two recent reports on young people 15–25 years of age in the UK (Livity, 2014) and 46 countries globally (Universum Global, 2014) concluded that a majority (65%) are generally optimistic and hopeful about the future, although this optimism does not necessarily apply to all aspects of employment (see below). A study titled *Life Patterns* conducted in Australia, considering two cohorts who left secondary school in 1991 (Cohort 1 – Generation X) and 2006 (Cohort 2 – Generation Y or Millennials), found that three priorities ranked much higher than a range of others: "to have a special relationship with someone"; "to have financial security"; and "to care and provide for a family" (Crofts et al., 2016).

There was evidence that the later cohort learned from the experiences of the earlier one and adjusted their expectations accordingly. For example, the lower importance given to financial security by Cohort 2 might have reflected a better understanding of the uncertainty of the labour market that Cohort 1 had experienced. Similarly, participants in Cohort 2 were less optimistic about having a secure, well-paid job within five years, and expected that it would take longer to develop a special personal relationship and establish a family. For both cohorts, the proportion who thought it was "very likely" they would own their own home in their early thirties was only around 30%. In-depth interviews indicated that both cohorts expected to rely heavily on financial support from parents or in-laws to meet their goals.

Yearly surveys between 2014 and 2017 of around 8,000 selected Millennials (the survey included only those who had a college or university degree, were employed full time and worked predominantly in large, private-sector organisations) concluded that: "In terms of overall confidence about the future, we observe general anxiety in mature markets and a more positive outlook in emerging markets" (Deloitte, 2017, p. 5). A moderate and slightly increasing level of economic optimism among this generation was reported in mature (developed) economies (increasing from 47% to 57% of participants between 2014 and 2017). However, the level of confidence among participants in emerging economies was both lower and more static (31% to 34%). In both types of economies, a new question in 2017 revealed low confidence that the social/political situation in their country would improve (48% and 25% in developed and emerging countries respectively). This lack of optimism regarding social progress was most evident in the following countries: South Korea, Mexico, Belgium, France, Chile, Germany, Japan, the UK, Australia and Italy. Greater optimism was seen in the Philippines, Peru, Brazil, India, Indonesia, Turkey, Argentina and Canada.

When asked to think about the world in general and how they feel about the future, in only 11 of the 30 countries covered did a majority expect to be happier than their parents. The perception that the previous generation enjoyed generally happier times was held most strongly in Japan, South Korea and countries in mainland Europe (including France and Germany). Millennials in India, Colombia, China, Peru, the Philippines and Indonesia were most convinced that they would be happier than their parents. These perceptions seem to relate to the issues that Millennials say are of concern to them: Terrorism (especially in northern Europe); conflict and political tension generally; crime/corruption and hunger/healthcare/inequality (more so in emerging countries); and unemployment.

Notably, these employed Millennials ranked climate change and resource scarcity, taken together, at the top of their list of concerns in the 2014 survey, but by 2017 environment/climate change ranked toward the bottom. This is in marked contrast to a 2018 study of 12–13-year-old school students in Austria and Australia, which found that a large majority of students in both countries believed strongly that climate change is something to worry and take action about *now* and believed this more strongly than adults in their respective countries (Harker-Schuch et al., 2018). These differences, both among the 2014–2017 Millennials surveys, and between the employed Millennials and school children in Austria and Australia, suggest that attitudes are likely to change for young people who find stable employment.

ATTITUDES AND ASPIRATIONS ABOUT EMPLOYMENT

Reports about young people's attitudes toward employment show commonality about what is hoped for. In the *Life Patterns* study of young Australians reported above, job security was a "very important" or "important" consideration for both cohorts when deciding on a career job at the age of 26–27 (Crofts et al., 2016). Similarly, in the UK, a majority of 16–25-year-olds in 2014 said they hoped for a rewarding career, most likely in a large corporation that practised responsible profit-making for the good of the world (Livity, 2014). They said they wanted to work for brands they admire, either "because they sell cool products, are recognisable and visible in the media or are run by people they consider as role models" (Livity, 2014, p. 5). Universum's survey of 50,000 15–19-year-olds across 46 countries in 2015 reported a desire to be themselves and express their personalities at work (Universum Global, 2014).

Some literature reports pessimism about achieving these hopes, whereas some expresses optimism. Nearly 40% of participants in the Universum survey feared they wouldn't get a job that would allow them to meet their aspirations. Beckett (2018) went further in his review of a range of literature, citing historian Benjamin Hunnicutt as saying that, in the USA, "belief in work is crumbling among people in their 20s and 30s" (Beckett, 2018, p. 1) and giving the example of the rise of "bullshit" (pointless and even socially damaging) jobs – drawing on the work of anthropologist David Graeber (2013). At the other extreme, the Deloitte survey of highly qualified, and mostly employed, Millennials reported that 76% regarded business as a force for positive social impact (Deloitte, 2017). Participants in this survey believed that employment in large enterprises gave them more potential to influence society than being in other jobs or unemployed. Flexible working conditions were cited as a key determinant of wellbeing, health and happiness (but only when accompanied by job security).

The Deloitte report concluded that criticism of business has diminished since 2015, with fewer participants saying that businesses focus on their own agendas rather than considering wider society. This positivity is tempered by the strong view that large enterprises are not fulfilling their potential to alleviate society's challenges, while charities and local, smaller companies are thought to be doing as much as they can. The perceived gap between actual contribution and potential contribution by large enterprises was greater in developed countries than emerging ones. Respondents thought that businesses could help particularly in relation to: education, skills and training; economic stability; cybersecurity; healthcare and disease prevention; unemployment; climate change; and unemployment. There was an even split between those who thought governments and business were working well together. Countries where Millennials believe government and business are working well together included the USA, the UK, Canada, Switzerland and Australia.

HOW LIFE TURNED OUT

There have been few studies over long time periods that investigated how life turned out for young people in comparison with their initial hopes and fears. Nevertheless, some trends emerge that seem likely to be relevant to most, if not all, generations. As discussed previously, surveys of young people around the world consistently reveal a strong desire for job security, full-time work, a healthy balance between work and private life, and fulfilling family and personal relationships. At the age of 24 to 25 (around seven years after leaving school) a little less than half of the Generation X and Y participants in the *Life Patterns* study in Australia said they were "very satisfied" with how life had turned out in relation to the hopes they had when leaving school, and only about 30% were "very satisfied" with their work and career status or accomplishments (Crofts et al., 2016). Around one quarter of the Generation Y cohort said they felt "mentally unhealthy" 10 years after first trying to enter the workforce. There was evidence that this generation had already begun to reduce their expectations based on the experiences of Generation X.

Employment and life experiences for both Generations X and Y in Australia were influenced by economic stresses early in their working lives – a recession for Generation X and the Global Financial Crisis for Generation Y: "Both generations have experienced changing labour market expectations that pressured them to gain skills and further education qualifications to navigate an evolved labour market" (Crofts et al., 2016, p. 8). In comparison, it is worth noting Elder's study of two cohorts of young men and women in the USA either born in the Great Depression or hoping to enter the workforce when the depression hit (Elder, 1998). That study found that financial deprivation, altered family relations and social/emotional strains had impacts early in life but that a large proportion of the study's participants recovered to lead happy and fulfilled lives. Our focus here, however, is not on how fulfilled future workers might be but whether or not they might choose occupational practice to seek that fulfilment. We return to this question after considering some implications of gender, below.

INFLUENCE OF GENDER ON HOW LIFE TURNED OUT

Gender is a major factor that can lead to inequalities among young people (Woodman & Wyn, 2015). Although there has been a dramatic increase in women's participation in education and in the workforce over the past few decades, many commentators over the past decade have raised concerns that this "gender revolution" is "unfinished", "incomplete" or even "stalled" (Cuervo & Wyn, 2011). It is beyond the scope of this chapter to focus on gender inequalities in detail, but two trends and a conclusion from the Australian *Life Patterns* study (Crofts et al., 2016) illustrate the complexity of gender issues in relation to aspirations and outcomes around work and life fulfilment. Among the participants who entered the workforce in the early 1990s, a progressively lower proportion of women than men were in full-time work as time went by (69% versus 79% at age 27 and 32% versus 90% at age 37). This appeared to be largely because women were more likely to take parental leave and/or return to work on only a part-time basis after giving birth. By the age of about 40, the gap was smaller although still large (68% versus 86%).

IMPLICATIONS FOR PRACTICE FUTURES

A key aim of this chapter is to explore how understanding past and present attitudes, hopes and fears of young people might help us anticipate future possibilities for occupational practice and, in particular, what types of future occupational practice might, or might not, wish, and be able, to attract and keep young recruits. We suggest that at least three conditions are necessary for occupational practice to be attractive to young people in the future: education, training and socialisation institutions and processes should exist that meet the business needs of professions and the occupational and personal needs of young people (Higgs, 2012); occupations should exist in forms that fulfil young people's lives; and, young people should believe that there are achievable life patterns that allow them to move from youth education (in whatever form that takes in the future) into occupational development that could proceed throughout working life.

In Chapter 3 of this book, Cork and Alford explored a range of plausible future scenarios for work and the occupations/professions, including ones that were characterised by combinations of community versus individual values and connected (interdependent) versus fragmented (autonomous) governance arrangements across society. Integrating education, training and socialisation institutions and processes to attract and prepare young people for futures that will demand multi, inter and even transdisciplinary knowledge and skills as well as advanced interpersonal and problem-solving skills would more likely be achieved in futures that have a focus on fostering intergenerational opportunity and integrated thinking and knowledge sharing across disciplines. This could be difficult to achieve in futures driven by individualistic values and disconnected governance within businesses and governments.

In the longer term, most future scenarios envisage increasing replacement of humans by artificial intelligence for many tasks, including many components of complex occupations that previously required humans to help society make sense of them. In the short to medium term, however, it seems highly likely that many occupations will require young people who can direct and manage intelligent machines that are performing routine and complicated tasks identified by human problem solvers and strategic thinkers. In many occupations, it is unlikely that there will be room for as many young people in the workforce as have been recruited in the past, leading to strong competition for positions. As shown by the fledgling *Making Our Future Work* project, the *Life Patterns* study and several other surveys around the world, as discussed above, young people are increasingly concerned that the opportunities to obtain these advanced skills will not exist in forms that they can afford. This will create long periods of financial and emotional struggle for them in their early to middle working years, exacerbated by delayed development of supportive personal relationships, and they are likely to be forced to take jobs that require less investment in education because they cannot wait long enough for opportunities in advanced occupations.

Recent surveys have shown that young people place high importance on having meaningful and ethical employment, and yet there is increasing evidence that many jobs, even within the more complex occupations, are becoming more routine and less meaningful – hence the rise of the term "bullshit jobs" (Beckett, 2018). It is also apparent that businesses, especially large corporations, are not living up to expectations and their potential to meet society's needs. Several business-led surveys of young people have highlighted the importance of businesses addressing the issue of job meaningfulness if they wish to attract and keep future employees. Of course, it is possible that future employers will not need to worry about labour shortages, but it is questionable whether the more complex occupations could survive if there was no incentive for lifelong learning and employee development.

Finally, the future of occupations is likely to be influenced by interactions among social, economic and/or environmental pressures, on the one hand, and the ways in which businesses, governments and society more broadly have prepared for, and respond to, these pressures, on the other. Those men and women who emerged fulfilled in later life after spending their early years living through the Great Depression attributed their fulfilment to: policies and programs that helped that generation obtain higher education; marriage as a source of critical support during their young adult years; and, especially, service in the military (in World War II and the Korean War) (Elder, 1998). Some of these opportunities allowed that generation to pursue careers in occupations requiring years of education and training, but others merely provided access to whatever jobs were available. Andres and Wyn (2010) investigated the influence of education and labour market policies on young people's capacities to start families in Canada and Australia. Australians took longer to secure a stable financial position that would enable them to form a family and support children. Similarly, the *Australian Oral History* project (Thomson, 2016) revealed the perceived importance to the Baby Boomers generation of the Whitlam Government's policy of free university in the 1970s, which allowed even financially disadvantaged young people to pursue tertiary education. We will not attempt to anticipate the types of economic ebbs and flows that Australia and the world might face in the future, except to note that the past two decades suggest that almost anything is possible.

Although there have been similarities in hopes and fears from generation to generation across the past century, there seems to be increasing pessimism among recent generations about future education, training and employment opportunities. In some possible futures these concerns might turn out to be valid, but in others the future could be better than young people expect. In all futures, however, it will be important for young people to avoid being overloaded with fears based on hearsay and untested opinions floating around society and to be able to think critically and strategically about the range of future possibilities and uncertainties so that they can build resilience and adaptability to shocks and be prepared to shape the future they want when opportunities arise. If we, as a society, want to help young people lead fulfilling lives, then society must provide opportunities for young people to build their resilience and future-thinking skills, or at least not hinder them from doing so.

Although there have been similarities in hopes and fears across generations from the past century, there seems to be increasing pessimism among recent generations about future education, training and employment opportunities. In some futures these concerns might turn out to be valid, but in others the future could be better than young people expect. In all futures, however, it will be important for young people to be able to think critically and strategically about the range of future possibilities and uncertainties so that they can build resilience and adaptability to shocks and be prepared to shape the future they want when opportunities arise. If we want as a society to help young people lead fulfilling lives, then society must generate opportunities to bolster this ability and preparedness. As noted in the Millennials Survey (Deloitte, 2017), businesses, in collaboration with governments, are in ideal positions to facilitate these sorts of arrangements for the future, and there is considerable potential that has not yet been tapped into.

NOTE

[1] http://australia21.org.au

REFERENCES

Andres, L., & Wyn, J. (2010). *The making of a generation.* Toronto, Canada: University of Toronto Press.

Beckett, A. (2018, January 19). Post-work: The radical idea of a world without jobs. *The Guardian.* Retrieved from https://www.theguardian.com/news/2018/jan/19/post-work-the-radical-idea-of-a-world-without-jobs

Brown, M. (1984). Young people and the future. *Educational Review, 36*(3), 303-315.

Crofts, J., Cuervo, H., Wyn, J., Woodman, D., Reade, J., Cahill, H., & Furlong, A. (2016). *Life patterns: Comparing the generations.* Melbourne, Australia: University of Melbourne.

Cuervo, H., & Wyn, J. (2011). *Rethinking youth transitions in Australia.* Melbourne, Australia: University of Melbourne.

Deloitte. (2017). *The 2017 Deloitte millennial survey*. London, England: Author. Retrieved from https://www2.deloitte.com/content/dam/Deloitte/global/Documents/About-Deloitte/gx-deloitte-millennial-survey-2017-executive-summary.pdf

Eckersley, R. (2008). Nihilism, fundamentalism, or activism: Three responses to fears of the apocalypse. *The Futurist, January-February*, 35-39.

Elder, G. H. (1998). *Children of the Great Depression: Social change in life experience* (25th anniversary ed.). United Kingdom: Avalon Publishing.

Foley, T., Hickie, I., Holmes, V., James, C. G., Rowe, M., & Tang, S. (2016). *Being well in the law: A guide for lawyers*. Retrieved from http://lawsociety.com.au/about/YoungLawyers/Publications/Beingwellinthelaw/index.htm

Graeber, D. (2013, August). On the phenomenon of bullshit jobs: A work rant. *Strike! Magazine*. Retrieved from https://strikemag.org/bullshit-jobs/

Harker-Schuch, I. E., Mills, F., Lade, S., & Colvin, R. (2018). Opinions of 12 to 13-year-olds in Austria and Australia on the worry, cause and imminence of Climate Change. *BioRxiv*. Retrieved from http://doi.org/10.1101/333237

Henley Centre. (1991). *Young eyes: Children's vision of the future environment*. London, England: Henley Centre for Forecasting.

Hicks, D. (1996). A lesson for the future: Young people's hopes and fears for tomorrow. *Futures, 28*, 1-13.

Higgs, J. (2012). Practice-based education: The practice-education-context-quality nexus. In J. Higgs, R. Barnett, S. Billett, M. Hutchings, & F. Trede (Eds.), *Practice-based education: Perspectives and strategies* (pp. 3-12). Rotterdam, The Netherlands: Sense Publishers.

Hutchinson, F. (1992). *Futures consciousness and the school: Explorations of broad and narrow literacies for the twenty-first century with particular reference to Australian young people*. Armidale, Australia: University of New England.

Johnson, L. (1987). Children's visions of the future. *The Futurist, 21*(3), 36-40.

Livity. (2014). *Young people in their own words*. Welwyn Garden City, England: Tesco.

Loughlin, C., & Barling, J. (2001). Young workers' work values, attitudes, and behaviours. *Journal of Occupational and Organizational Psychology, 74*(4), 543-558.

Malbon, J. (2017). *Making our future work*. Canberra, Australia: Australia21. Retrieved from http://australia21.org.au/wp-content/uploads/2017/12/Australia21-Making-our-Future-Work-Pilot-Workshop-report.pdf

McCrindle, M., & Wolfinger, E. (2009). *The ABC of XYZ*. Sydney, Australia: UNSW Press.

Thomson, P. A. (2016). Australian generations? Memory, oral history and generational identity in postwar Australia. *Australian Historical Studies, 47*(1), 41-57.

Universum Global. (2014). *Generation Z grows up*. Retrieved from https://universumglobal.com/insights/generation-z-grows/

Woodman, D., & Wyn, J. (2015). *Youth and generation: Rethinking change and inequality in the lives of young people*. London, England: Sage Books.

Steven Cork PhD (ORCID: https://orcid.org/0000-0002-3270-4585)
Crawford School of Public Policy
Australian National University, Australia
Ecoinsights, Australia
Australia21, Australia

Jennifer Malbon B. Interdisciplinary Studies (Sustainability) (Hons.)
UNSW Canberra

JAMES CLOUTMAN AND GRAHAM JENKINS

16. FACING RECRUITMENT CHALLENGES

Entering Workplace Practices

When I was very young, I didn't have any work. I went and stayed with friends in the country for a year or two, and I didn't work until my father said, 'it's about time you came home and did something'. I went into town, and I had three interviews. And by the time I got on the tram and went back to Kingsford my father said there were two phone calls for me and there were two jobs being offered. Those interviews were very short, and I had no references because I hadn't worked before. And in those days, those particular days, your education was more important than anything you'd done, and that was all people wanted to know. (Research participant recalling applying for work in the 1960s, Cloutman, 2018, n.p.)

INTRODUCTION

The world of work has changed dramatically in the last few decades, particularly in the last 20 years with the advent of shorter-term contract arrangements, ICT technology, new working arrangements in the "gig economy"[1] and the emergence of social media (Brown, Hesketh, & Williams, 2004; Brown, Lauder, & Ashton, 2011; Withers, Lewis, & Dalton, 2016). Other chapters in this book discuss in depth the implications of these changes for workers and employers, including the type of employees that employers are seeking. This chapter deals with staff recruitment.

The challenge of recruiting employees, particularly when highly skilled workers or many staff are required for busy organisations, has increasingly led employers to use recruitment companies. In recent decades the role of recruitment companies has expanded across appointment levels and range of jobs in relation to helping employers gain suitable employees. Recruitment companies have radically changed their structure and approach in many respects during this period to keep up with industry's transformations. Some recruitment companies have taken over the recruitment responsibilities of organisations.

This chapter investigates this changing world of modern recruitment and how best employers and (prospective) employees can interact with these changes. The chapter begins with a brief history of the recruitment industry, with how it emerged and the changes that have taken place in the industry, particularly in recent years. Next, we deal with how recruitment companies operate today, the various types of recruitment companies and the reasons why large organisations may utilise their services. Lastly, the chapter reviews how potential clients might optimally utilise them.

© KONINKLIJKE BRILL NV, LEIDEN, 2019 | DOI: 9789004400795_016

A BRIEF HISTORY OF THE RECRUITMENT INDUSTRY IN AUSTRALIA

Emerging during and immediately after World War II, the recruitment industry in Australia began by finding workers for the agricultural sector, which had been badly affected by labour shortages as men left to join the armed forces. Over the next three decades, with the advent of the manufacturing sector, the recruitment industry expanded beyond agriculture and gradually began to provide staff to a range of industries, particularly the tertiary (service) sector. Shopfront recruitment centres, often called "temp agencies" or "personnel agencies", began to emerge in the sixties, providing employees for lower-paid, often administrative positions in the fast-growing major cities.

With the emergence of a modern economy and outsourcing in the 1980s, the demands on recruitment companies expanded to include the sourcing of all levels of staff up to and including CEOs and board members. This evolution was accompanied by the emergence of a range of different types of recruiters, each focused on a specific level or category of employee. Today, as industry continues to transform, ICT technology and social media have presented yet further challenges to the recruitment industry as their clients have tapped into these technological resources to directly contact potential staff, possibly circumventing recruitment intermediaries altogether (KPMG, 2016). Nevertheless, last year the recruitment industry gained over $3 billion in revenue and had around 7,300 companies in Australia, employing 22,500 staff. And the industry continues to expand.[2]

HOW RECRUITMENT COMPANIES OPERATE

For anyone approaching the job market for the first time and even for those who've already been in it for several years, some of the typical questions that people have about recruiters might include: What exactly are recruitment companies and what do they do? Are they all similar and if there are many different types of recruiters how do you know which ones can aid you and how do you design a strategy for interaction with them? Is the recruitment company's client the employer or employee?

Recruitment companies are employer-oriented. They sit between employers, the job applicant, and the hiring entity, the company or organisation that is seeking staff. Recruitment companies are not there to find jobs for workers but exist to serve organisations, their clients. Recruiters are, in the main, a type of intermediary salesperson selling their services to organisational clients. While these companies may be promoting a role to their candidates (prospective workers), and appear to be working on the candidates' behalf, it is typically the organisation who has paid the fee and any commitment to candidates does not bring obligation to find them work.

Essentially, recruiters are intermediaries who take over some of the work of the human resources (HR) team in their client companies by matching jobs with suitable applicants, checking that they have the skills and experience that a business or organisation requires. Recruiters identify applicants for both permanent and temporary positions and, although recruitment companies vary considerably in type, they all work in roughly the same way, following the processes outlined below.

- An organisation seeking staff provides a description of the relevant role(s) to the recruiter, outlining the level of seniority, indicative salary and required skills and qualifications needed for a position.
- The recruiter then uses a variety of techniques to identify potential candidates (applicants) for their client (the hiring organisation). These techniques can include a search within the recruiter's internal database, searches within online CV databases (such as that provided by SEEK), online and/or in-print advertising, social media searches and recommendations made through networks.
- The recruiter compiles a "long list" of potential candidates and then starts approaching and interviewing them. Interviews may be face-to-face or they may be by telephone. Typically, the recruiter will be attempting to evaluate the candidate's track-record and skills against the selection criteria for a job as provided by the client company. They will also be assessing motivation to leave a current employer, salary expectations and overall fit to the culture within the client's organisation.
- Based on these interviews, the recruiter prepares a "short list" of suitable candidates and sends it to their client or reviews it with their client face-to-face.
- Clients then select those from among the short list that they wish to interview. The recruiter working on this assignment then arranges interviews, evaluations such as psychometric testing, carries out reference checks, coaches candidates through the resignation process with their current employer, helps them deal with any counter-offers, places the candidate in the new job and stays in touch with them during the probation period to ensure that there have been no "on-boarding" issues. The recruitment company may also provide a guarantee, typically three to six months for more senior positions, to replace the candidate if they do not work out in the role for any reason.

THE DIFFERENT TYPES OF RECRUITMENT COMPANIES

While the processes outlined above are typically followed by most recruiters, they may vary with the type of recruitment company and the nature of the role that they are seeking to fill. There are several different types of recruitment organisations. These include the following.

JobActive[3]

This is a free recruitment service provided by the Australian government for jobseekers (Fowkes, 2011). Any person wishing to use this service should register first with Centrelink,[4] who will assign the person to a specific provider: either a JobActive agency or, in the case of people with disability, a Job Access[5] provider. JobActive services are currently delivered through 1,700 sites nationwide by 44 private sector – charitable and for-profit – providers. No sites are government-owned. The emphasis within the JobActive network is on locating suitable work as quickly as possible so that prolonged unemployment is avoided.

Labour Hire Recruitment Companies

What distinguishes recruiters in this category is that they act, essentially, as the de facto employer of candidates/workers, hiring and payrolling them and providing workers compensation and other statutory cover as well as other benefits, even though the employee is, typically, working on-site with the client company. Recent years have seen a significant rise in labour hire recruiters locating staff for clients and directly contracting them, often as self-employed contractors who carry their own insurances as a Pty Ltd company (Connelly & Gallagher, 2004; International Labor Organization, 2015). Labour hire recruiters typically specialise in one of the following four areas, though these are fluid, rather than rigid, categorisations:

- Recruiters working with candidates who may be seeking work in either the industrial or agricultural sectors. Typical roles would include more junior to mid-level positions in mining, construction, logistics, transportation, warehousing, agriculture or food processing. Recruiters in this category may also source candidates for administrative or management positions within these sectors though this may not necessarily be their main focus. Examples of such labour hire companies include Workpac[6] and Programmed.[7]
- Specialist recruiters working to place staff in sectors such as the health or catering sector who may also be seeking shorter-term contract or casual staff to meet the organisation's needs. Recruiters in this category work with candidates seeking roles in nursing, aged care, disability care, community service and commercial catering. Recruiters in this category also often work with candidates seeking FIFO (fly-in, fly-out) roles for staff providing catering and food services in remote locations such as offshore platforms, mining stations and military bases. Examples include Pulse[8] and Pinnacle People.[9]
- Recruiters providing staff for the IT sector for such roles as programmer, data analyst, desktop support, project manager or change manager (Barley & Kunda, 2011). Examples include Greythorn[10] and PeopleBank.[11]
- Recruiters working to provide white collar, administrative staff who are also, typically, hired on a casual or contract basis. Specialising in lower level roles up to, for example, $80,000 per annum, these types of recruiters seek administrative, clerical, office support and reception staff. Examples of such recruitment companies are Kennedy Reid[12] or Adecco.[13]

Executive Recruitment Companies

Executive recruiters work to fill management and executive level roles from, for example, from $100,000 per annum in total remuneration all the way up to "C" level roles such as CIO and CEO and board positions. There are many types of recruiters working in this category from smaller, niche recruitment organisations filling roles in one industry sector to larger, more generalist executive recruitment companies who predominantly handle white collar management level roles (see Table 16.1). This category of recruiter also includes the "elite" level of recruitment companies who work only on the most senior executive positions such as those required by ASX listed companies. For these higher-level roles such as C level positions or executive directors, organisations may use the services of an executive search firm. Such recruiters usually work on a retained basis, rather like a lawyer, and will only engage with their clients if they are retained. These companies, such as Heidrick and Struggles[14] or Egon Zehnder,[15] often started in management consulting in the 1950s. They quickly realised that their clients had a need for assistance in recruiting the most senior – and often the hardest to hire – staff and, over the decades since their inception, have developed a deep expertise and specialist knowledge of the fields in which their clients operate.

As a general rule, labour hire companies are seeking to build a business by contracting manpower out to their client companies. In contrast, executive search firms seek to generate revenue through placing candidates in permanent roles at their client companies in exchange for a placement fee. However, these categorisations are only indicative and there is often overlap between these different types of company. Fees charged for these services are as varied as the types of recruiters. Fees for labour hire are, typically, in the 6–15% of salary range while fees for executive search vary from 12% for lower level roles all the way up to 35–40% for the most senior positions.

Graduate Recruitment Specialists

University graduates could consider any of the types of recruitment organisations outlined above, depending on their studies, career goals and previous work experience. Graduates who have had no work experience to date could benefit from seeking out recruitment companies specialising in graduate recruitment, such as Grad Australia[16] or Chandler McCleod.[17]

Larger organisations may have ongoing "staff matching" strategies such as intern programs that these specialist recruiters may be able to access on graduates' behalf. During these internships new employees are involved in professional development that is organisation as well as self-beneficial, with further matching of employees (through streamlining and "weeding-out" processes).

WHY – AND HOW – DO ORGANISATIONS USE RECRUITERS?

Years ago, you could go into a recruitment consultancy and could be assured of work within a week or two. I've gone to recruitment companies earlier on in my career, and before I got home, there was a message [from them]. Bringing it fast forward through to today ... recruitment consultants leave people in despair. I don't blame the recruitment consultants; it's the market they are in at the moment. When recruitment consultancies advertise online, they can't cope with the response. It's a tsunami of applicants. They too, like any other business, they've cut back on their staff – they will only pick out the ones that they want to pick out. (Cloutman, 2018)

Table 16.1. Types of Australian recruitment companies and their roles.

Type of recruiter	Primary function	How do they aid the individual worker?	How are they remunerated?	How do you find them?	How do you encounter them?
JobActive Consultants	To assist unemployed people find employment via government programs managed by Centrelink & the Dept. of Employment	Assistance depends on contractual obligations between the service provider and the Dept. of Employment	By the federal government	By registering with Centrelink	Face-to-face through a structured engagement process
Labour Hire Recruitment Consultants	To assist corporate clients find suitable interim or project-based employees that the recruitment company payrolls	Focus is on aiding their fee-paying corporate clients; however, as the relation with candidates is a contractual one a longer-term relation may ensue	By their clients	By submitting resumes and possibly by direct approach	Face-to-face at their invitation
Executive Recruitment Consultants	To assist corporate clients find suitable employees (at all levels)	Highly limited. Their focus is on aiding their fee-paying corporate clients	By their clients	By submitting resumes and possibly by direct approach	Face-to-face at their invitation

One question that many people may have regarding recruitment companies is why would organisations use them when they have their own HR departments, often with a recruitment section? Would they not be better off doing recruitment themselves?

However, even though larger corporations do, indeed, attempt to do much of their own recruiting there are many reasons why organisations use intermediaries for recruiting of new staff. These include:

- **Cost reduction:** Organisations may be able to lower their overall manpower costs through outsourcing. Additionally, by using a recruiter with extensive knowledge of a specific market sector, they may be fast-tracking the hiring of staff and filling otherwise empty seats, thus reducing lost productivity.
- **Focus:** It may be the case that an organisation doesn't wish to deal with hiring and prefers to concentrate on other aspects of their business. Recently, for example, the sheer volume of responses to advertising may lead organisations to conduct a cost-benefit analysis of their hiring processes and conclude that they would be better off outsourcing the handling of recruitment to an intermediary.

- *Discretion:* In some cases, for example, there may be someone internally that senior staff are planning to terminate and do not wish to let the market know of such a strategy, which would be the case if they were to advertise to fill the role directly. Organisations may also be seeking to alert a competitor to a new initiative they are undertaking by "camouflaging" the hiring process via an executive search firm.
- *"Try before you buy":* By employing someone via a recruitment company through a short-term contractual arrangement – that is later extended to full-time employment – employers may be testing the fit of a candidate, ensuring that they truly have the skills being sought and that they fit well into the corporate culture.
- *Accessing specialised skills:* Some roles are so highly specialised or in such demand that an organisation knows, often through trial and error, that they are likely to fail or to have a very hard time finding suitable staff. In such an eventuality, it makes more sense to work via a specialist recruiter who may already have suitable candidates identified than to attempt to find them directly.
- *Professional expertise:* It may simply be the case that the skills of a highly professional recruitment organisation that is constantly approaching and sourcing candidates and continually asking for referrals outstrips the ability of an organisation's own HR department to find suitable staff. Recruitment is often a form of sales in that recruiters are "selling" a job to candidates who may need persuasion to leave their current employment and HR staff may lack such skills or be unwilling to develop them.
- *Previous attempts to hire have failed:* Lastly, it may also be the case that, despite attempts to hire internally and through advertising, no suitable candidates have made it through the selection process. In this case a hiring organisation may have little choice but to use a recruitment company or face the prospect of other staff leaving because of the burdensome workload in a particular unit or department.

Historically, many large organisations had recruiters that they preferred to work with and, in the main, many organisations left hiring up to line managers and their personal preferences until the early 1990s. However, these relations then started to be formalised to a greater degree through preferred supplier arrangements (PSAs), largely as a result of the recession in the early part of that decade and the need to cut costs. In the years following the recession this trend continued with companies continually reviewing costs and often bringing in procurement specialists whose job it was to negotiate PSAs with recruiters. In some cases, particularly in the industrial recruiting sector, fees dropped as low as 6–8%, from a high of 40% in some cases.

Aggressive reduction of margins has had a number of ramifications for the recruitment industry and many recruiters have had to restructure their operations as a result. For example, some recruitment companies have transitioned from having a team of recruiters handling every aspect of the hiring process to a structure whereby a director or team leader delegates the work of candidate research and screening to more junior support staff while handling the final stages of the recruitment process themselves. Unfortunately, other, often seasoned recruiters have left the industry, choosing instead to take roles in clients' internal recruitment teams. These cost-cutting and streamlining measures in the recruitment industry – along with increased levels of competition for jobs in recent years – have sometimes led to the sense on the part of applicants for jobs that working with a recruiter may not always be a positive experience.

USING A RECRUITER TO FIND EMPLOYMENT

I've had a good experience [with recruiters], but not everyone does have a good experience. I've never had a bad experience with one. I have had non-replying and people that don't get back to me and all that sort of stuff, but I think, now I know [from] being in the industry, sometimes it's impossible to stay on top of things. (Cloutman, 2018)

[The recruitment industry] is fuelled by a lot of younger people who actually don't have much business experience because they come into it in their early twenties. They also are very driven by sales because that's what it is, it's a very dog-eat-dog world. And, because of that, they are driven to acting in behaviours which work in the short-term but not in the long-term. Candidates feel like they are just fodder sucked through in order to meet their [targets]. (Cloutman, 2018)

Working with recruitment companies can, for many people, be a mixed experience ranging from "being ignored" to "being a second priority when compared to the employer" to "gaining positive feedback and a good job". The differences in experience largely relate to the expectations that a candidate may have of the recruitment industry to start with and, critically, to their understanding of how the recruitment industry operates and how best to approach it so as to maximise outcomes. Some useful advice for people using recruitment companies is as follows:

- Remember that recruiters fill only about 15% of Australia's jobs; the rest are filled by companies themselves so focus on direct approaches to employers. These direct approaches need to be driven by a "goal-oriented, dynamic, self-regulatory [job searching] process" (Van Hoye & Van Hooft, 2013, p. 3).
- Note that large companies have multiple ways of recruiting. While they may use recruitment firms, they may also advertise, may have employee referral programs and may approach candidates directly themselves through LinkedIn or via information saved in databases of past employees or applicants that the HR department maintains (Gerard, 2011). Find out the typical recruitment practices in your industry, including how and where advertisements appear and whether recruitment agencies are used.
- Since both recruiters and the organisations they serve are increasingly utilising social media and ICT technology to nurture potential employees and source talent, explore social media when job searching and maintain a credible social media presence yourself rather than just a CV. This is now critical in many instances to job acquisition success.

- With regard to ICT technology, bear in mind that many large recruitment companies or corporates use computerised "applicant tracking" software, such as JobVite, that sifts and sorts bulk applications by key words. Unless a resume has the same key words in it that have been programmed into the tracking software being used to manage the hiring process for a particular role, even highly suitable candidates' resumes could be passed over. It is critical to continually monitor job sites such as SEEK and LinkedIn and ensure that your resume reflects the most up-to-date words used to depict roles in your sector.
- Plan your career trajectory and consider when you can benefit from career advice and help through mentors and recruitment sites and companies.
- Make sure that your CV is up-to-date and suits (e.g. through key words) the job(s) you're applying for. Keep resumes to 2–3 pages and have a longer resume available on request.
- Look for relevant advertisements and/or engage with the recruitment companies directly to optimise interaction and avoid mistaken inferences by recruiters based solely on a CV (Cole et al., 2008).

Top candidates often take an over-the-horizon approach to dealing with the recruitment industry. Having compiled a list of relevant recruiters in their sector, these candidates stay in touch with them via social media, sending the recruiters regular updates on their activities and inviting them to conferences where they are speaking. Some candidates will also, if the opportunity presents itself, use selected recruitment companies to build a long-term personal relationship that they know may be of use to them in the longer-term when they are looking for work.

A point of reflection here: the practices and advice above may also be relevant to job searching within an organisation for internal job moves. Recruitment departments in larger organisations have adopted many of the practices displayed by recruitment companies and, for those already in employment, a prudent, longer-term job searching strategy most likely includes selectively employing techniques outlined above in both intra-organisational and extra-organisational job search.

WHERE TO FROM HERE FOR THE RECRUITMENT INDUSTRY?

It can be seen from the information provided in this chapter that the history and development of recruitment companies tend to follow both the fortunes of, and the transformations taking place in, the broader corporate and industry worlds. Given market turbulence in the last two decades, hiring freezes, steep declines in recruiters' fees, and the ever-increasing use by larger organisations of internal recruitment teams within their HR departments it might seem strange that recruitment companies are still thriving.

Yet, even despite these challenges, the recruitment industry has survived and is still expanding. Perhaps there will always be a need for people-focused intermediaries who can solve the challenges of finding talent but there will also undoubtedly continue to be change within both the recruitment industry and the larger corporate/industry world that it services. Here are some of the key areas where further change can be expected:

- *Technological transformation:* With downward pressure on fees and increased pressure to succeed in challenging markets, recruitment companies will continue to seek out new ways to expedite processes through the utilisation of technology. This may include "chatbots" to aid in screening candidates, greater use of AI)-based evaluation of candidates to speed up assessment and more intense utilisation of software to both generate job ads and sift and sort applications. These influences may well make engagement with recruitment companies seem more transactional and tech-driven, rather than the person-to-person interaction that was a feature of engaging with recruiters in the past.
- *Diversity:* As corporations seek to deepen their customer insight and acquire new markets, recruiters will be increasingly tasked with hiring diverse labour groups including more mature employees and applicants from among ethnic minorities and the disadvantaged.
- *New forms of interviewing:* Video or online interviewing, virtual reality assessments/gaming-based evaluations and job "auditions" (rather than interviews) exist in the present and may be increasing trends of the future (Deloitte Global Human Capital Trends, 2017). Potential candidates will be tasked with a short period of "test work" so that they can be observed – whether by the company directly or through an intermediary such as a recruiter – and evaluated on the job.
- *A de-emphasis on traditional qualifications:* Recruiters are likely to be increasingly tasked with identifying the soft or generic skills that are so highly valued by today's organisations, rather than with only checking for traditional certifications such as a university degree (LinkedIn Talent Solutions, 2018). Addressing this goal, universities are increasingly offering options for students to gain micro credentials as well as expanded graduate learning outcomes.

CONCLUSION

This chapter has outlined transformations in recent decades in worker recruitment and, in particular, the recruitment industry. Both recruiters and the organisations that they serve have undergone and are likely still to undergo significant transformation in the way that they hire and retain employees with advanced, technology-driven techniques for hiring coming to the fore. In a rapidly transforming world of work recruiters are still an active intermediary force because the personal touch, market knowledge and expertise will always be valued by the corporations that they serve. Within larger organisations recruitment departments are adopting similar advanced strategies to the recruitment companies.

NOTES

[1] This term refers to environments where temporary positions are common and organisations contract with independent workers for short-term engagements.

[2] For specific details on the recruitment industry's financial performance and other relevant statistics please refer to IBIS World at https://www.ibisworld.com.au/
[3] https://jobactive.gov.au
[4] An Australian Government, Department of Human Services agency: https://www.humanservices.gov.au/individuals/centrelink
[5] https://www.jobaccess.gov.au/
[6] https://www.workpac.com/
[7] https://programmed.com.au
[8] https://www.pulsejobs.com/australia/
[9] https://www.pinnaclepeople.com.au/
[10] https://www.greythorn.com.au/
[11] https://www.peoplebank.com.au/
[12] https://www.kennedyreid.com.au/
[13] https://www.adecco.com.au/
[14] https://www.heidrick.com/
[15] https://www.egonzehnder.com/
[16] https://gradaustralia.com.au
[17] https://www.chandlermacleod.com/graduate-recruitment

REFERENCES

Barley, S., & Kunda, G. (2011). *Gurus, hired guns, and warm bodies: Itinerant experts in a knowledge economy*. Princeton, NJ: Princeton University Press.

Brown, P., Hesketh, A., & Williams, S. (2004). *The mismanagement of talent: Employability and jobs in the knowledge economy*. Oxford, England: Oxford University Press.

Brown, P., Lauder, H., & Ashton, D. (2011). *The global auction. The broken promises of education, jobs and income*. New York, NY: Oxford University Press.

Cloutman, J. (2018). *Towards a new learning paradigm in employability for mature-aged workers in the professional and business (tertiary industry) sectors (proposed title)* (PhD in preparation). Charles Sturt University, Wagga Wagga, Australia.

Cole, M., Feild, H., Giles, W., & Harris, S. (2008). Recruiters' inferences of applicant personality based on resume screening: Do paper people have a personality? *Journal of Business and Psychology, 24*(1), 5-18.

Connelly, C., & Gallagher, D. (2004). Emerging trends in contingent work research. *Journal of Management, 30*(6), 959-983.

Deloitte Global Human Capital Trends. (2017). *Rewriting the rules for the digital age*. Retrieved from https://www2.deloitte.com/insights/us/en/focus/human-capital-trends/2017/future-workforce-changing-nature-of-work.html

Fowkes, L. (2011). *Rethinking Australia's employment services*. Sydney, Australia: The Whitlam Institute, University of Western Sydney.

Gerard, J. (2011). Linking in with LinkedIn®: Three exercises that enhance professional social networking and career building. *Journal of Management Education, 36*(6), 866-897.

International Labor Organization (ILO). (2015). *World employment social outlook: The changing nature of jobs*. Geneva, Switzerland: Author.

KPMG. (2016). *The Australian recruitment industry: A comparison of service delivery 2016*. Retrieved from https://www.jobs.gov.au/news/australian-recruitment-industry-comparison-service-delivery-2016

LinkedIn Talent Solutions. (2018, January 10). *Global recruiting trends 2018: The four ideas changing how you hire* [Web log post]. Retrieved from https://business.linkedin.com/talent-solutions/blog/trends-and-research/2018/4-trends-shaping-the-future-of-hiring

Van Hoye, G., & Van Hooft, E. (2013, August). *New directions in understanding job search: A self-regulatory perspective* [Conference abstract]. Paper presented at the Academy of Management Annual Meeting, Lake Buena Vista, FL.

Withers, G., Lewis, P., & Dalton, T. (2016). *Report on casual and part time work in Australia*. Canberra, Australia: Fair Work Commission.

James Cloutman MEd (ORCID: https://orcid.org/0000-0001-9907-5632)
PhD Candidate, Charles Sturt University, Australia
Member, Education, Practice and Employability Network, Australia

Graham Jenkins
Director, Graham Jenkins Pty Ltd

BERNADINE VAN GRAMBERG

17. PHDS AND FUTURE PRACTICE

It has long been assumed that getting a Doctor of Philosophy (PhD) is essential for an academic career. However, in recent years only a small fraction of PhD graduates secure an uninterrupted academic career. In Australia over 60% of PhD graduates take up positions in industry and some have hybrid careers between academia and industry. This trend is seen elsewhere in the world and has led to significant changes in the training and development of PhD students as well as initiating changes in the preparation of academics who supervise them. This chapter explores this trend towards diverse PhD graduate destinations and examines how universities are preparing PhD students for future practice internationally. In doing so, the chapter describes traditional PhD training and the growth and nature of non-academic destinations for PhD graduates. Finally, the chapter investigates how universities are preparing PhD graduates for future practice including understanding the characteristics of the future of work, and it illustrates this with a case study of the evolution of PhD training at Swinburne University of Technology.

THE TRADITIONAL PHD

The PhD is almost universally considered to be the highest university award attainable. The award emerged in the middle ages along with master's degrees and indicated that a high level of teaching and scholarship expertise had been achieved (Daly, 1961). The more modern development of the award as a research qualification emerged in 19th century Europe where the *philosophiae doctor* was named after the "love of wisdom" (*philosophiae*) and "to teach" (*doctor*) (Oldfield, 2011). At this time the production of a research thesis, or dissertation, became evidence of scholars' ability to learn by enquiry and marked their area of specialisation. The intent was to create a class of professors who would educate future members of the profession and teach into specialised fields (Lash, 1987).

In line with this intent, training PhD students has traditionally been conducted under the supervision of a panel of PhD holders and taken the form of an academic apprenticeship where students are taught to become independent researchers, attain the skills and tools to undertake good research, and learn how to teach (Park, 2005). In Australia, the PhD is described as producing graduates who "have systematic and critical understanding of a complex field of learning and specialised research skills for the advancement of learning and/or for professional practice" (Australian Qualifications Framework Council, 2018, p. 63). This is reflected in the reasons many graduates cite for moving into academia. For instance, Bexley, Arkoudis, and James' (2013) survey of academics found that the key reasons for working in a university were: the opportunity for intellectually stimulating work (95.9%); passion for a field of study (93.8%); and the opportunity to contribute to new knowledge (91.1%).

The Australian PhD degree typically takes between three to four years of independent (supervised) research generally culminating in a thesis which is examined by experts outside of a student's university (Australian Council of Graduate Research, 2015). Over time, it has become an expectation that for an academic career one needs to hold a PhD. That said, it has been well known for more than a decade that a significant proportion of PhD graduates take up jobs outside academia and I consider this next.

THE SHIFTING CAREER PATHS OF PHD GRADUATES

Australian data on PhD graduate employment destinations demonstrates that there has been quite a shift from predominantly university-based employment to industry or government. For instance, a study by Western et al. (2007) on PhD graduate destinations from the Group of Eight (Go8) universities found that only half of the doctoral students surveyed worked in the higher education sector. Similarly, Neumann and Tan's (2011) longitudinal study of PhD holders who graduated between 2000 to 2007 found that graduates had a high rate of employment (90% were in work), with 44.4% employed in the higher education sector – but of this group only 54.9% were in academic positions.

Four key drivers for the diversification of PhD careers are considered here. First, there has been a rapid growth in the numbers of PhD graduates over the past two decades. A recent report on the changing PhD by the Go8 noted that the number of doctoral enrolments grew by 68% between 2000 to 2007 and the number of completions almost doubled per year, resulting in more graduates than academic positions available in Australia (Go8, 2013).

Second, in Australia, undertaking a PhD reflects an ethic of lifelong learning in the sense that commencing PhD students are on average about 33 years old and more than 10% of commencing students are over 50 years of age (Go8, 2013). Students come to their studies from a range of working backgrounds. Many (58.9%) work part time while they complete their PhD (Edwards, Bexley, & Richardson, 2011). Most of this employment is with their universities but 25.8% work outside the university sector. PhD students not only come from a variety of occupations prior to their PhD studies, but they also have a variety of expectations of their future career after their PhD. For instance, Edwards et al. (2011) note that while 62.8% of respondents intended an academic career, 54.1% considered this ambition as a medium to long-term prospect, aiming to do something different eventually. Additionally, 41.9% of respondents had intentions to work outside the university sector either in professional or research-related fields (Edwards et al., 2011).

Third, decreased government funding and budget pressures in Australian universities have led to changes in the way universities employ academics. May, Peetz, and Strachan (2013) report that since the 1980s academics have increasingly been employed on a casual basis, particularly at lower academic levels. This has led to fewer, ongoing jobs for new PhD graduates. Indeed, Hugo (2008) refers to this as representing a "lost generation" of potential academic staff due to a combination of the failure to hire and the success of these graduates securing work in other areas of the economy (including going offshore for work). Not only are there fewer jobs in academia, but many of the traditional benefits of this work – "tenured positions: autonomy; a balance of research, teaching and service; job security (perhaps a job for life), and clear and largely linear career pathways" (Bexley et al., 2013, p. 397) – have been eroded, making academia less desirable for many graduates as a career.

Fourth, PhD careers have diversified over the years because the emergence of the knowledge economy has driven changes in the nature of work and the skills and knowledge required to perform that work. Neuman and Tan (2011) argue that a knowledge-based economy is reliant on a supply of research-trained employees who "effectively combine highly specialised research and industrial and economic capacity building" (p. 602). Others have added that universities are the key drivers of the knowledge-based economy (Department of Innovation, Industry, Science and Research [DIISR], 2011; NOUS Group, 2017).

Internationally, the same trends are observed. A set of studies by Nerad et al. (2007) in the US found that 5 to 10 years after their PhD graduation, less than half of the graduates worked in universities. Similarly, the 2016 report by the League of European Universities found that "many graduates from research training courses go into roles beyond research and education, in the public, charitable and private sectors, where deep rigorous analysis is required" (Bogle, 2014, p. 6).

CHANGES TO DOCTORAL EDUCATION

These changes in employment patterns of PhD graduates have implications for doctoral education. Recognising this, in 2011 the Australian Federal Government laid down its research workforce strategy to 2020. The report notes that "[t]his vision recognises the central role of our human capital – in particular, individuals with HDR qualifications and specialised technical skills – in strengthening both our capacity to innovate and our ability to adopt or adapt innovations developed elsewhere" (DIISR, 2011, p. 9). Its recommendations included that the training of Higher Degree by Research (HDR) students must include "not only academic skills, but a wide range of generic competencies to operate effectively in these diverse contexts" (p. 22). Similarly, in 2013, the Go8's discussion paper on the changing PhD drew on international innovations such as embedding industry work placements as part of the PhD and creating Doctoral Training Centres which bring students together in groups and particularly linking them to industry. This was followed by a report by the Australian Government Department of Education (2014) which recommended a set of initiatives to increase industry engagement opportunities for research students, including industry internships and professional development courses.

In 2015 the Australian Federal Government commissioned the Australian Council of Learned Academies (ACOLA) to investigate the current state of research training for HDR students. Their report (McGagh et al., 2016) indicated that overall Australian PhD holders are well trained and highly competent in their chosen fields. However, they noted that the model for PhD training still relies on an academic apprenticeship even though the greater proportion of PhD graduates find work in industries other than education. The report recommended sweeping changes in the training and preparation of PhD graduates for work outside the education sector. These included formalising coursework and practical training for PhDs to provide transferrable skills for employability in all industries; increasing industry-university collaboration in the supervision of PhDs; encouraging industry placements; benchmarking PhD training against world standards; and professionalising research supervision.

The ACOLA report recommendations have been echoed by others. For instance, in 2016, the Australian Technology Network (ATN) of universities, a cluster of five technologically specialised institutions, commissioned a report on research training and its relevance to industry (NOUS Group, 2017). The report pitched the value to industry of PhD students and graduates and recommended that: deeper links need to be made between industry and universities; PhD employability needs to be targeted; and a communication strategy to improve awareness of the utility of research-trained employees is needed along with tax incentives for business to engage PhD graduates and students.

These various reports have been greatly influenced by developments particularly in the UK and Europe. I now briefly consider this horizon scanning before turning to the changes being made to research training in Australian universities.

AN INTERNATIONAL HORIZON SCAN OF RESEARCH TRAINING

The idea behind changing PhD training to meet industry needs is certainly not new. As Australia was commencing the debate on how to make PhD training more relevant for those moving into industry jobs, such changes were already being implemented across the world. The Roberts report (2002) sparked the introduction of professional development training in the UK and later informed the Researcher Development Framework (Vitae, 2011). The Researcher Development Framework is the most prevalent set of research training benchmarks for universities in the UK (Bogle, 2014). It consists of 12 categories and is supported by an online tool through which academics and research students can measure their performance and development needs across both academic and industry skill sets.

In 2005 the European Union endorsed the Salzburg Principles for research training across Europe's universities. These 10 principles focused on the third cycle of the Bologna process – the doctoral training years. Briefly, these principles cover the importance of undertaking original research in the thesis, and encourage institutional strategies to ensure professional career development opportunities, diversification of doctoral programs, strengthening of supervision skills and capability, consistency of candidature duration, student mobility and sustainable funding (Christensen, 2005).

In 2011 the European Commission added to this a checklist of seven Principles for Innovative Doctoral Education in a bid to bring about a baseline for quality and consistency for all students in EU Higher Education Institutes. Based on the Salzburg principles, the seven benchmarks facilitate an environment for research exchanges across European universities and industry partners, joint PhDs and shared supervision. This was seen as vital as applications for doctoral education had been increasing dramatically and these future PhD holders were envisaged to lead innovations in industry across the world (ERA Steering Group on Human Resources and Mobility, 2015). The benchmarks also emphasise institutional capacity, interdisciplinarity and industry engagement (see also Bogle, 2014).

In implementing the principles European universities have focused on enhancing industry engagement for doctoral students. Two key reports include *Research Careers in Europe* (Scholtz et al., 2010), which recommended that researcher development encompasses whole of lifecycle research career planning with greater connection between universities and businesses, and the European University Association's *Doctoral Education* (2016), which focused on PhD models with industry co-supervision to build a knowledge-based economy. I now turn to how these measures have influenced the developments in PhD training in Australia.

TOWARDS AN INDUSTRY-ENGAGED AUSTRALIAN PHD

As noted earlier, the ACOLA report (McGagh et al., 2016) into research training has been a major trigger for reconsidering how universities prepare their PhD students. However, Australia faces a major obstacle – the current lack of collaboration between universities and industry, and perhaps a reticence on the part of industry to engage with university research. The ACOLA report listed concerns that Australia does not effectively translate its research into societal or economic benefits; that there is insufficient knowledge transfer between researchers and industry; and that Australia has lower levels of commercialisation of research than other countries. In its discussion paper on the engagement of universities with industry, the Australian Academy of Science asserted that "Australia ranks 29th out of 30 for industry-university collaboration in the OECD" (Early- and Mid-Career Researcher [EMCR] Forum 2016, p. 2) and that industry is more likely to source innovation from a competitor or other business in the same industry than from a university. This was reiterated in the NOUS Group (2017) report which found that industry engagement with research students is low and recommended that "[a] national communication strategy led by Government and supported by universities and peak bodies, to highlight such skill development amongst PhD holders, may help improve awareness and understanding of PhDs" (p. 38).

By way of addressing the low levels of industry engagement, the ACOLA report made some specific, industry-focused recommendations, including coursework which features practical transferrable skills training for employability; increasing industry-university collaboration in the supervision of PhDs; and encouraging industry placements. At the same time as the ACOLA report was commissioned, the federal government also commissioned an investigation into research policy and funding arrangements (Watt, 2015). The Watt review was aimed at finding opportunities to strengthen collaboration between universities and businesses, and its recommendations included to provide funding to increase incentives for universities and businesses to engage in research; for universities to report on their industry engagement activities commencing 2018; and to provide funding for industry placements for PhD students.

CASE STUDY: THE TRANSFORMATION OF THE PHD AT SWINBURNE UNIVERSITY OF TECHNOLOGY

In response to the recommendations of the ACOLA report and other government and international recommendations as outlined in this chapter, Australian universities are changing the way PhDs are delivered. In this section I outline the response of Swinburne University of Technology in developing new models of doctoral training.

Swinburne University of Technology

Swinburne is a multi-sector provider that offers higher education, vocational education and pathways, and TAFE courses and programs, along with online education in partnership with Open Universities Australia and through Swinburne Online. The university has multiple campuses in Melbourne, Australia, and also an international campus in Sarawak, Malaysia. The university has around 55,000 enrolled students including 1,200 PhD students.

Swinburne has been ranked in the world's top 400 universities by the *Times Higher Education* World University Rankings and 61 amongst universities less than 50 years old in the Young University Rankings (Swinburne University of Technology, 2017a). In keeping with its original mission, the university's emphasis is on high quality, engaged teaching and research in science, technology, business, design and innovation – teaching and research that makes a difference in the lives of individuals and contributes to national economic and social objectives (Swinburne University of Technology, 2017b).

The Changing PhD at Swinburne

The challenges of engaging industry with doctoral training have been met in two ways at Swinburne. First, two doctoral programs specifically train students in an industry setting. The Industry PhD embeds students in the workplace where they work on a three-year research plan which provides research and development (R&D) solutions for the business and trains students to translate their research findings into business applications. The Industry PhD is jointly supervised by staff from Swinburne and industry and has 15 students currently enrolled working in hospitals and technology firms. In contrast, the Practice-Based PhD is designed for senior industry practitioners who focus on researching a business problem reflecting on their years of experience. This is a cohort-based model which provides research training and sharing of industry and university expertise. There are currently 67 students enrolled from government and private firms including Victoria Police, the CSIRO and management consultancies.

Second, while all Australian universities are moving towards greater coursework components in their PhDs which offer professional development and transferrable skills, Swinburne provides an embedded Graduate Certificate (Research and Innovation Management) which is separately awarded. The program has been designed to enhance the experience, career prospects and employability of PhD students. The primary aim of the program is to support the development of industry-ready applied researchers and leaders across all fields.

The Graduate Certificate has two compulsory units comprising Innovation and Creativity in Research and Project Management for Research. By including a compulsory unit in innovation, the Graduate Certificate contributes both to the university's strategic vision and to the development of knowledge economy practitioners. The unit on Research Project Management focuses on transferrable skills in managing the research process, budgets, timelines, stakeholders and risks. Figure 17.1 outlines the structure of the Graduate Certificate (Research and Innovation Management), depicting the core units and some of the possible elective units. Both core units contribute to the development of a student's research proposal; early feedback from students suggests that the training may help them achieve a timely completion of their degree. Two elective units allow students to strengthen their knowledge in a number of areas including research methods, professional writing or to take a work placement in industry.

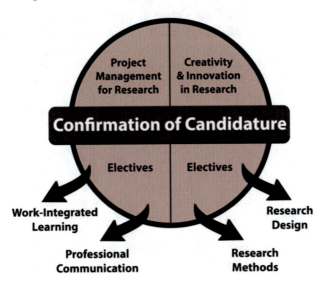

Figure 17.1. Depiction of the Graduate Certificate (Research and Innovation Management) at Swinburne.

Encouraging Industry Placements

In setting up the Graduate Certificate at Swinburne, it was important to draw upon external and internal resources. There are a range of external options for securing research work placements in Australia. First, bodies such as the Australian Mathematical Science Institute (AMSI, 2018), the Business Higher Education Round Table (BHERT, 2018) and Ribit (2018) provide options including work placements, conferencing and online matching of students with a range of participating employers. Whilst still in its early days, this indicates a changing industry practice with regard to R&D in Australia (Apri Intern, 2018; Ribit, 2018).

In November 2017, Swinburne and Ribit collaborated to run the first PhD Speed Network event in Australia which brought together over 100 PhD students with 35 employers (Ribit, 2017). Employers included government departments, major banks, management placement firms, technology firms and transport service firms. The final year students were asked to register with Ribit and to prepare a six-minute introduction outlining their key research skills. At the Speed Network evening, Ribit's Spin Doctor was available to listen to the students' six-minute pitches and to help them focus on their transferrable skills. Feedback has been positive from students and employers, most of whom indicated an interest in attending a second event. Thus far no placements have been recorded although applications are currently being considered by employers. Currently, five students are being interviewed for positions and nine applications are in progress.

CONCLUSION

This chapter has canvassed the shift in employment destinations for Australia's PhD graduates which has implications for doctoral training and supervisory practices. Universities such as Swinburne have responded rapidly to the opportunity to maximise their engagement with industry in order to drive the knowledge economy. This has included developing new models of doctoral training which provide students with industry-embedded experience; as well as coursework. The major challenge in scaling up industry-focused PhD training to meet the future needs of Australia's PhD students is the lack of awareness and willingness of many Australian businesses to engage with universities, their research staff and students. That said, future practice in the knowledge economy is likely to lead to large numbers of industry positions which require research qualifications. Rather than becoming a relic of the past, the PhD is already being transformed into a mechanism for businesses to innovate through employing skilled and talented workers who are capable of independent and original research and development. It is likely employer reticence will be a short-term problem simply because of the vast numbers of highly competitive PhD graduates entering the employment market looking for opportunities to use their skills.

REFERENCES

Apri Intern. (2018). *Transform your business with an Australian Post Graduate internship.* Retrieved from http://aprintern.org.au/
Australian Council of Graduate Research. (2015). *Graduate research good practice principles.* Retrieved from https://www.ddogs.edu.au/good-practice-principles
Australian Government Department of Education. (2014). *Initiatives to enhance the professional development of research students.* Canberra, Australia: Commonwealth of Australia. Retrieved from https://docs.education.gov.au/system/files/doc/other/initiatives_to_enhance_the_prof_development_of_research_students_0.pdf
Australian Mathematical Sciences Institute (AMSI). (2018). *About us.* Retrieved from https://amsi.org.au/about-us/
Australian Qualifications Framework Council. (2013). *Australian Qualifications Framework Second Edition.* Retrieved from https://www.aqf.edu.au/sites/aqf/files/aqf-2nd-edition-january-2013.pdf
Bexley, E., Arkoudis, S., & James, R. (2013). The motivations, values and future plans of Australian academics. *Higher Education, 65*(2), 385-400.
Bogle, D. (2014). *Good practice elements in doctoral training* (Advice Paper 15). League of European Universities. Retrieved from http://www.ub.edu/escola_doctorat/sites/default/files/internacionalitzacio/LERU_good_practice_doctoral_training_final.pdf
Business Higher Education Round Table (BHERT). (2018) *About BHERT.* Retrieved from https://www.bhert.com/about
Christensen, K. K. (2005). *Bologna Seminar on doctoral programmes for the European knowledge society* (General Rapporteur's Report). Salzburg, Austria. Retrieved from http://www.eua.be/eua/jsp/en/upload/Salzburg_Report_final.1129817011146.pdf
Daly, L. J. (1961). *The medieval university: 1200–1400.* New York, NY: Sheed and Ward.
Department of Innovation, Industry, Science and Research (DIISR). (2011). *Research skills for an innovative future: A research workforce strategy to cover the decade from 2020 and beyond.* Canberra, Australia: Commonwealth of Australia. Retrieved from https://docs.education.gov.au/system/files/doc/other/research_skills_for_an_innovative_future.pdf
Early- and Mid-Career Researcher (EMCR) Forum. (2016). *Starting the conversation between academia and industry consultation* (Discussion Paper: Consultation Draft, September). Retrieved from https://www.science.org.au/files/userfiles/support/emcr-activities/academia-industry-discussion-paper-consulation-draft.pdf
Edwards, D., Bexley, E., & Richardson, S. (2011). *Regenerating the academic workforce: The careers, intentions and motivations of higher degree research students in Australia: Findings of the National Research Student Survey (NRSS).* Melbourne, Australia: Centre for the Study of Higher Education, Australian Council for Educational Research. Retrieved from https://research.acer.edu.au/higher_education/23
ERA Steering Group on Human Resources and Mobility (ERA SGHRM). (2015). *Using the Principles for Innovative Doctoral Training as a tool for guiding reforms of doctoral education in Europe.* Retrieved from https://cdn5.euraxess.org/sites/default/files/principles_for_innovative_doctoral_training.pdf
European University Association. (2016). *Doctoral education – taking Salzburg forward: Implementation and new challenges.* Brussels, Belgium: Author. Retrieved from http://www.eua.be/Libraries/publications-homepage-list/Doctoral-Education_Taking-Salzburg-Forward.pdf?sfvrsn=18
Group of Eight (Go8). (2013). *The changing PhD* (Discussion Paper). Retrieved from https://go8.edu.au/sites/default/files/docs/the-changing-phd_final.pdf
Hugo, G. (2008). *The demographic outlook for Australian universities' academic staff* (CHASS Occasional Paper 6). Adelaide, Australia: Council for Humanities, Arts and Social Sciences.

Lash, L. L. (1987). The nature of the Doctor of Philosophy degree: Evolving conceptions. *Journal of Professional Nursing, 3*(2), 92-101.

May, R., Peetz, D., & Strachan, G. (2013). The casual academic workforce and labour segmentation in Australia. *Labour and Industry, 23*(3), 258-275.

McGagh, J., Marsh, H., Western, M., Thomas, P., Hastings, A., Mihailova, M., & Wenham, M. (2016). *Review of Australia's research training system* (Report for the Australian Council of Learned Academies). Melbourne, Australia: ACOLA. Retrieved from https://acola.org.au/wp/PDF/SAF13/SAF13%20RTS%20report.pdf

Nerad, M., Rudd, E., Morrison, E., & Picciano, J. (2007). *Social science PhDs – five+ years out: A national survey of PhDs in six fields* (Highlights Report). Seattle, WA: Centre for Innovation and Research in Graduate Education. University of Washington. Retrieved from http://www.education.uw.edu/cirge/wp-content/uploads/2012/11/ss5-highlights-report.pdf

Neumann, R., & Tan, K. K (2011). From PhD to initial employment: The doctorate in a knowledge economy. *Studies in Higher Education, 36*(5), 601-614.

NOUS Group. (2017). *Enhancing the value of PhDs to Australian industry*. Canberra, Australia: Australian Technology Network of Universities. Retrieved from https://www.atn.edu.au/siteassets/publications/atn01-phd-report-web-single.pdf

Oldfield, S. (2011). The doctor of nursing practice graduate and the use of the title 'Doctor'. In H. M. Dreher & M. E. Smith Glasgow (Eds.), *Role development for doctoral advanced nursing practice* (1st ed., pp. 369-384). New York, NY: Springer.

Park, C. (2005). New variant PhD: The changing nature of the doctorate in the UK. *Journal of Higher Education Policy and Management, 27*(2), 189-207.

Ribit. (2017). *PhD speed dating for jobs @ Swinburne University, Melbourne* (23/11/2017). Retrieved from https://www.ribit.net/blog/event-news/20171123-phd-speed-dating-jobs-swinburne/

Ribit. (2018). *What is Ribit?* Retrieved from https://www.ribit.net/about-us/

Roberts, Sir G. (2002). *SET for success: The supply of people with science, technology, engineering and mathematics skills*. London, England: HM Treasury. Retrieved from http://webarchive.nationalarchives.gov.uk/+/http://www.hm-treasury.gov.uk/d/robertsreview_introch1.pdf

Scholz, B., Vuorio, E., Matuschek, S., & Cameron, I. (2010). *Research careers in Europe: Landscape and horizons* (A Report by the ESF Member Organisation Forum on Research Careers). France: European Science Foundation. Retrieved from http://archives.esf.org/fileadmin/Public_documents/Publications/moforum_research_careers.pdf

Swinburne University of Technology. (2017a). *Strategic Plan 2025*. Retrieved from https://www.swinburne.edu.au/media/swinburneeduau/about-swinburne/docs/pdfs/Swinburne-Strategic-Plan-2025.pdf

Swinburne University of Technology. (2017b). *Times Higher Education ranks Swinburne as a top young university*. Retrieved from http://www.swinburne.edu.au/news/latest-news/2017/04/times-higher-education-ranks-swinburne-as-a-top-young-university.php

Vitae. (2011). *Vita Researcher Development Framework 2011*. Retrieved from https://www.vitae.ac.uk/vitae-publications/rdf-related/researcher-development-framework-rdf-vitae.pdf

Watt, I. (2015). Review *of research policy and funding arrangements*. Canberra, Australia: Commonwealth of Australia. Retrieved from https://docs.education.gov.au/system/files/doc/other/main_report_final_20160112.pdf

Western, M., Boreham, P., Kubler, M., Laffan, W., Western, J., Lawson, A., & Clague, D. (2007). *PhD graduates 5 to 7 years out: Employment outcomes, job attributes and the quality of research training* (Final Report, Revised). Brisbane, Australia: The University of Queensland Social Research Centre (UQSRC).

Bernadine Van Gramberg PhD, Swinburne University of Technology, Australia

ASHELEY JONES

18. EDUCATIONAL INNOVATIONS

Preparing for Future Work

The world of work has changed significantly over the last few decades and the education sector, in its current form, is struggling to maintain the pace of change required to meet industry needs. In light of the swift transition to a digital era, it is imperative to concentrate attention on a better alignment of educational training to career pathways that are relevant to the emerging trends informing the global employment marketplace. This chapter aims to explore a model of educational advancement that, through the effective disruption of assessment within current forms of educational practices, may offer an innovative way to better prepare graduates for future work requirements.

There is little doubt that employability in the 21st century requires people to be adaptable, lifelong learners with a desire for engaged learning. Yet the role universities should play in the production of such work-ready graduates has long been contested. The types of skills that graduands should possess, outside of discipline-specific knowledge, has been a matter of discussion for many decades across and within academic circles, professional associations, and industry and business forums. The traditional tiered education model, whereby Australian school leavers wishing to pursue further educational opportunities are asked to choose between formal award higher education programs or competency-based vocational education and training (VET) training packages, has maintained its dominance within the majority of professional labour development sectors. Educational models have not fundamentally changed in the last 100 years. This, in spite of the continuous move towards knowledge-based economies that require dexterous human capital cohorts equipped to meet the ongoing demands of the various employment markets.

Over the last few years, the debate regarding the relevancy of the current educational model has progressed to encompass the idea that whilst more young people than ever before are going to university, the value of the investment in this particular educational pathway, especially for millennials, is increasingly being brought into question (Bowles, 2017; Dede, 2017; Ernst & Young, 2012).

In light of the substantial increase in the award of tertiary qualifications, the occupational value of degrees has in turn declined. In much the same way that a high school leaving certificate was once deemed sufficient to enter clerical positions that led to junior management pathways, jobs that previously mandated undergraduate degree qualifications have now been inflated to require postgraduate qualification attainment. And yet, in spite of this increase in qualification requirements, employers continue to articulate concerns as to the quality of the graduate skill sets to which they have access.

There are many who suggest it is time to determine a better alignment of our educational institution offerings to encourage improved diversity in training opportunities so as to create better career paths for early careerists, or those seeking to reskill (see, for example, World Economic Forum, 2016). It is important to acknowledge that mid-career retraining will become ever more essential as the skill mix required for effective career changes increases. Ensuring these groups receive the pertinent skills to apply in the workforce, through a flexible model that concentrates on the development, or enhancement of skill sets, is increasingly important. The current education system focuses attention on individual accomplishment; yet, attributes such as collaboration, communication and conflict resolution have fast become the seminal skills required in a global workforce dominated by complex situations necessitating cross-disciplinary initiatives.

Ernst & Young (EY) (2012) in their report *University of the Future* go so far as to interrogate the viability of current university modelling. The writers of this report predict that within 5–10 years the existing model will no longer be sustainable in light of the move by millennials away from traditional university block mode teaching. Similar sentiments are captured in the Productivity Commission's 2017 report, *Shifting the Dial: 5 Year Productivity Review*, which states that there is a clear need for

> an efficient, high-quality and flexible education and training system that is driven by the needs of users (the people acquiring the skills and the businesses that need them) rather than the interest of suppliers or legacy models of provision and government funding. That system also needs to be able to respond to the inevitable transitions from job to job and occupation to occupation that will occur over people's lifetimes. (p. 84)

The report goes on to argue:

> the current skills system has fractures that put at risk its capacity to deal with the future labour market changes. There are deteriorating results among school students. The VET system is in a mess and is struggling to deliver relevant competency-based qualifications sought by industry. Leading segments of the university sector are more focused on producing research than improving student outcomes through higher-quality teaching. (p. 86)

Regardless of whether one agrees or disagrees with these criticisms, there is undoubtedly a need to reimagine the current education model. This is made more apparent when considering the data provided within a further 2017 report published by the McKinsey Global Institute (Manyika, 2017), which contends that:

> **by 2030, 75 million to 375 million workers (3 to 14 percent of the global workforce) will need to switch occupational categories** [bold added]. Moreover, all workers will need to adapt … Some of that adaptation will require higher educational attainment or spending more time on activities that

require social and emotional skills, creativity, high-level cognitive capabilities and other skills relatively hard to automate. (p. vii)

REVIEWING THE LEGACY MODEL

Accepting that disruption within the current education model has not fundamentally occurred, it is important to note that whilst there is a clear need for a review of the current legacy model, this call to arms is in no way intended to throw the baby out with the bath water. The current suite of four-year undergraduate and two-year master's programs undoubtedly have a place within the educational ecosystem. It is perhaps the "vanilla flavouring" of such courses that is one of the areas that needs to be interrogated. For example, is there really a need to offer the same information technology (IT) and nursing courses across institutions, rather than have innovative, agile offerings that permit a point of difference in the specialisms offered to those investing significant time and money in these training programs? In order for university programs and VET training packages to maintain sufficient currency to keep apace of the fast-changing requirements of the global economy, the bureaucracy surrounding changes in program offerings needs to be far more flexible and agile in order to provide relevant training within the economies of the 21st century. The revision and development of curricula offerings, for example, can no longer be subject to lengthy review processes that can take anything upwards of 12–24 months to implement. There is a need to provide labour-relevant qualification choices that meet the needs of the fast-paced industry changes that inform current workforce requirements.

This is a criticism that can equally be laid at the door of the VET training system, where training packages take an inordinate amount of time to develop and consequently fail to deliver in an agile fashion to the needs of employer requirements. At the behest of the Australian Government Department of Education and Training, Bowles (2017) conducted a review of existing VET training packages to determine current levels of automation training. The report revealed a paucity of units of competency addressing training in this area, in spite of the continued increase in automation-enabled economic growth requirements. The report contends that in order to remain competitive in a global economy "the focus has to shift to future employability and growing industry competitiveness, not just ensuring a person is competent to perform in a current job role" (Bowles, 2017, p. 3).

Yet, whilst the review identifies the importance of partnership alignments between industry and educational institutions to provide practical arenas – or internships – in which to apply technical skills and build the soft skills required to operate effectively within the workplace, the resulting strategy fails to identify a fundamental element; that is, a structured assessment measurement that can clearly determine competency across all internship graduate programs. The conspicuous absence of quality assurance regarding the efficacy of national internship outcomes would appear a matter of concern in terms of assessing the holistic success of such programs. There are no clear overarching measures to determine the successful application of the varying learning outcomes across institutions. Placing this requirement into context, we need only draw again from EY's 2012 report which identifies that for universities to survive and thrive, they will need to build significantly deeper relationships with industry in the coming decade. Scale and depth of industry-based learning and internships, for example, will become increasingly critical as a source of competitive advantage for those universities who have the industry partnerships and pedagogy to do it well.

A further potential cause for concern in the delivery of most internship components includes the acknowledgement that intern supervisors are not required to have any particular experience or undertake any specific training in the role of intern supervision. This potential inability to provide constructive, supportable student feedback from industry supervisors should be viewed as problematic. Given these types of programs are intended to ensure the work readiness of its graduates for transition into the workplace, there is a very real need to measure and publicly disseminate their success, if only in order to better inform other work integrated learning (WIL) models and ensure continuous improvement of current programs (Australian Collaborative Education Network [ACEN]), 2015).

It is not only the discipline-discrete knowledge, its delivery and the lack of structured assessment measurements that also need to be reimagined. Numerous reports produced across academic and business cohorts over the last few decades have suggested that graduates entering the workforce, as well as those requiring reskilling, are insufficiently equipped to meet employer requirements. For example, the World Economic Forum 2016 report, *The Future of Jobs*, analysed emerging job profiles and advertisements, confirming that 70% of job profiles for future workers are composed of non-technical skills. In much the same way that digital literacy has become an integral skill requirement across the employment spectrum, so too are the "softer" skills, such as communication, emotional judgement, creativity, critical thinking and problem solving. A 2017 Deloitte Access Economics report stated that "ten of the sixteen crucial proficiencies in the 21st century identified by the World Economic Forum are non-technical" (p. 1).

Building on the identification of future attributes required by those entering the workforce, Christopher Dede, Professor in Learning Technologies at Harvard University's Graduate School of Education conjectures "success a decade after high school graduation in a global, innovation-centred world will be as much determined by students' character and their ability to work with others as by their intellectual capabilities" (Dede, 2017, n.p.). Citing a 2012 report by the National Research Council, Dede goes on to posit that "a combination of cognitive, intrapersonal, and interpersonal skills—flexibility, creativity, initiative, innovation, intellectual openness, collaboration, leadership, and conflict resolution—are essential for keeping up in the 21st century. I would argue that instead of preparing students for careers, we should focus on inculcating skills that are transferable across many roles" (2017, n.p.).

IDENTIFYING AN OBJECTIVE ACCREDITATION SYSTEM

Reflecting on the issues outlined above, it is clear that genuine models of disruption of the current higher education models are required. These new models must not only embrace emerging forms of learning, but also transform the very nature of the educative landscape.

In order for these ideas to achieve traction within the education sector, there is a need to move focus away from the concept of traditional modes of teaching and assessing. Instead, attention needs to be focused on the development and implementation of experientially based, industry-endorsed, peer-assessed skill capabilities, which can be verified and delivered across a spectrum of horizontal skill sets. An exploration of the potential role micro credentialing could play in establishing iterative, assessable disruption of the current status quo offers one way to achieve such traction in disrupting and transforming the education landscape. A particular strength of the micro-credentialing approach is the ability to ensure both academia and industry partnerships are implemented that ensure all stakeholders shoulder the responsibility for innovative reform of the present education sector.

MEASURING WORKFORCE SKILLS CAPABILITIES

Bowles and Lanyon (2016) define a credential as "formal recognition by an authoritative, independent third party that has assessed and judged that a person through experience and workplace performance has achieved a particular professional capability" (p. 2). Accepting this definition of credentialing, it is important to note that this form of skills capability measurement model requires the use of authoritative and independent third-party assessment of a candidate's experiential application of pre-determined skill sets. As a result, assessment of this type is invariably reliant on the partnering of industry and academia to determine best practice iterative assessment processes that meet objective third-party pre-determined outcomes. Thus, a strength of the credentialing model is the ability of industry and academia to assess and award formal recognition of a candidate's experience and performance within the workplace.

Both digital badging and micro-credentialing practices are already well established. For instance, in 2013–2014 alone, institutions in the US awarded nearly 1,000,000 non-degree education awards and certificates. In Australia, the development of similar awards is gaining traction, with numerous universities, VET and private providers considering the role such practices might play in their own educational offerings, particularly in the areas of soft skill development or experiential skill recognition.

Based on the premise of competency-based capability recognition, micro credentialing has the ability to alter education on many levels. Foremost perhaps is the opportunity for industry, governments, professional associations and educational institutions to work together in the development of pertinent credentials that validate a candidate's skill sets in specific domains across agreed and measurable criterion. Utilising the experience of professional associations in the development and recognition of benchmarked, objective criterion to underpin skill relevancies across varying professions ensures credential models remain pertinent and cater to the ever-changing needs of workforce requirements.

Agility is the name of the game. The mapping of standards – agreed by industry advisory boards, similar to those conducted within higher education curriculum development reviews – that have been built to reflect capability across a structured, scaffolded credentialing process allows for a flexible, yet rigorous process to formally recognise an individual's talent capabilities. The ability to conduct benchmarked assessment against industry requirements is a powerful argument, particularly given the agility of skill development or skill procurement that will be required across the global workforce. As a result of the development of such structured benchmarking opportunities, companies will be able to develop customised in-house professional development training, which caters to targeted skills development across all staffing levels. Consequently, professional development training becomes outcomes focused, and targeted to the specific needs of a business unit or team, rather than an ad hoc approach to improving staff productivity.

This form of skill recognition offers candidates the opportunity to demonstrate capability in skills for which they do not necessarily hold formal award qualifications. This opens educative recognition pathways to a significant proportion of the population who previously did not qualify for formal award recognition. The very core of credentialed skills recognition is that it is experientially based and often evidence agnostic. This means it is proving capability of the skill set itself through varying evidentiary artefacts that is required, rather than traditional formative or summative assessment tasks that disadvantage all bar the most academically inclined. Credentialing of an individual's specific skill sets ensures an opportunity to build a global passport of endorsed capabilities, which have been verified by an independent assurance model. A second benefit is one of economic advantage to those seeking skill endorsement. The ability, for instance, to obtain credentials at a significantly reduced cost to that of formal award subject units, which can then offer building blocks to articulate into formal award courses, ensures educational attainment becomes an attractive option for the majority of the population.

Reviewing a Viable Credentialing Model

One of the most robust and mature examples of the development of a microcredentialing model is that initiated by DeakinDigital, the commercial arm of Deakin University, which is now known as DeakinCo.[1] This credentialing model is underpinned by professional capability standards that have been developed in collaboration with various professional associations and industry experts. These standards have then been mapped against the university's eight graduate attributes: the first involving discipline-specific knowledge and capabilities as mandated by the relevant professional accrediting body; with the remaining seven mapped to generic skill sets, including digital literacy, communication, teamwork, problem solving, global citizenship, self-management and emotional judgement.

EDUCATIONAL INNOVATIONS

Underpinned by the university's brand and reputation, Deakin University's Professional Practice credentials validate, and award, varying skill sets of candidates that better inform the capabilities of the organisations and professions within which they operate.

In order to meet the varying professional development requirements across individual capabilities, each of the credentials are scaffolded against varying levels of the Australian Qualifications Framework (AQF). This results in credential offerings across five levels of capability, from entry level (AQF3) to an advanced level (AQF 9) (see Table 18.1) – the latter offering formal pathways into potential professional practice master's degrees in the areas of IT, financial planning, leadership and digital learning.

Table 18.1. Deakin University Professional Practice credentials: Comparative alignment of levels.[2]

Deakin Professional Practice credentials	Bachelor-aligned	Pre Master's-aligned	Master's-aligned
How credential levels align to international qualifications frameworks			
Australian Qualifications Framework (AQF)	AQF Level 7 Bachelor degree	AQF Level 8 Graduate certificate Graduate diploma Bachelor Honours degree	AQF Level 9 Master's degree
European Qualifications Framework (EQF)[a] with Bologna Cycles	Level 6 – Cycle 1	Level 7 – Cycle 2	
Framework for higher education qualifications in England, Wales and Northern Ireland (FHEQ)[b]	FHEQ Level 6 Bachelor's degree Bachelor's degree with honours Graduate diploma Graduate certificate	FHEQ Level 7 Master's degree Postgraduate diploma Postgraduate certificate	
USA Higher Qualifications Profile (DQP)[c]	Bachelor's degree		Master's degree
How credential levels align to levels of work			
	Operational	Functional	Strategic

[a] EQF alignment based on research and analysis by the joint European Qualifications Framework, New Zealand Qualifications Authority and Australian Qualification Framework Council reported in the Australian Qualifications Framework (2014), Alignment of the AQF with the New Zealand Qualifications Framework and the European Qualifications Framework consultation paper.
[b] FHEQ alignment reference: http://www.qaa.ac.uk/docs/qaa/quality-code/bologna-process-in-he.pdf
[c] USA DQP alignment reference: https://www.luminafoundation.org/files/resources/dqp.pdf

215

Collaborating with Industry Associations

As previously indicated, one of the strengths of credentialing professional skills is the integral role professional associations can play in the benchmarking of agreed standards across levels of capabilities. Thus, an important component of the Deakin University model is the close working relationship maintained across various professional associations so as to ensure the provision of relevant industry subject matter and expert insights into the development of cutting-edge professional capability standards. Such types of partnering arrangements are specifically intended to inform the standards that determine the skills required in the Industrial 4.0 era. For instance, in early 2017, a cluster of digital marketing credentials was developed in collaboration with the Association for Data-driven Marketing and Advertising (ADMA). As a result of several roundtable discussions with senior marketers from both agency and commercial entities, the skill areas of Digital, Data-driven, Content and Creative Marketing were identified as pivotal for the development of future workforce capability across the professional marketing ecosystem. The huge increase in the uptake of data-driven and digital expertise across industry verticals has resulted in the need to assess and validate those working within this fast-paced environment. The development of such credentials was specifically intended for those professionals who do not have formal recognition for the skills they possess that are currently in high demand.

A further example of professional association input into the identification and development of a pertinent credential to meet current market needs was engagement with members of the Institute of Analytical Professionals of Australia (IAPA). Members of the IAPA community contributed significant subject matter expertise in the production of professional practice standards to inform a Data Analytics credential intended to benchmark skill capabilities within the fast-growing data analytics field. The determination of the skill sets required at each of the five levels of capability, across industry verticals, aligned to levels of autonomy, influence and complexity, ensures that careerists operating at all levels of data analysis are able to evaluate the skills they need to develop in order to attain the next level of credential. Similarly, those wishing to reskill or re-enter the workforce may also determine which skills they need to develop or hone so as to attain skill validation within this emerging area of professional expertise.

In much the same way that mature capability frameworks, such as the Skills Framework for the Information Age (SFIA), are used to determine skills within the IT industry; through evidence-agnostic determination, the Deakin University Professional Practice credentials operate in much the same way. The proof of evidence required to demonstrate capability of a skill level remains in and of itself entirely evidence agnostic. In other words, a candidate will be assessed for proving specific project management skills, rather than the ability to provide mastery of a specific project management tool or methodology.

Reviewing the Assessment Process

As indicated earlier the validity of such models is heavily reliant on the model's ability to provide an objective, iterative assessment methodology: one that affords each candidate a fair and equitable assessment experience. Such an assessment methodology needs to be repeatable across myriad sets of cohorts delivering similar results, irrespective of the choice of assessor.

Since those awarded Deakin University Professional Practice credentials may choose to use these awards to articulate into various formal award courses, the precision with which professional practice credential assessment is undertaken is rigorous. All assessments for credentials are evidence-based and evidence is determined from within a work-based application that provides currency of skill application, which is both reliable and provable.

For example, in each instance, two assessors are appointed to blind review a candidate's reflective testimony and uploaded evidence, along with a supporting asynchronous video testimony that requires response to a series of randomly generated questions, aligned to the relevant rules of evidence. In keeping with the Tertiary Education Quality and Standards Agency (TEQSA) requirements, credential assessors are required to hold a qualification one level higher than the potential level of the qualification an assessed candidate could obtain. The strength of the Deakin model is that the assessment design is a further collaboration between industry and academia in the determination of a candidate's satisfaction of the requirements for bestowal of the credential award. Thus, a candidate is assessed by both an industry peer as well as an academic assessor, ensuring the award of credentials meets both industry and academic standards. In the event that the two assessors vary in their assessment of a candidate's capability to satisfy the award of a credential, a third independent assessor will be appointed to review the evidence artefacts and determine a final assessment outcome.

At each stage of the process, candidates are exposed to transparent and iterative assessment processes that are also subject to similar governance and compliance requirements requisite across formal higher education award programs. Thus, the Deakin model can be seen to be agile in order to meet the needs of industry, whilst at the same time adhering to the stringent compliance requirements Deakin University is required to maintain in order to meet TEQSA standards.

CONCLUSION

The first section of this chapter identified the need for a better alignment of current educational institutional offerings so as to disrupt the identification and assessment of an individual's skill set, outside of contemporary summative assessment practices. The chapter went on to argue there is a very real requirement to design and implement authentic new ways to recognise industry specific skill requirements that can be benchmarked against agreed capabilities. An examination of the Deakin credential model was then undertaken as a viable case study to illustrate a robust and mature credentialing model. This was intended to offer an example of how traditional, well-established dual sector educational institutions such as Deakin University and DeakinCo can offer educational innovations through working with professional associations such as ADMA and IAPA in the design and implementation of innovative work-based assessment models that can better prepare the future workforce. As can be seen, this is achieved through an alignment of professional skill sets to an agile and flexible benchmarking of an individual's current skill sets. The ability to design new and pertinent evaluation structures that remain aligned to traditional formal award models highlights a way in which higher education and VET providers can reinvent current modes of assessment. This change in assessment practices is an essential requirement for educational institutions in order that they may become more dynamic and responsive to the needs of industry within a 21st century global workplace – a workplace that is reliant on agility and flexibility in order to meet the demands of continuous digital disruption.

NOTES

[1] Originally known as DeakinDigital, this entity has now merged with DeakinPrime (the Registered Training Organisation component of the Deakin stable) to become DeakinCo, trading as an arm of Deakin University.

[2] Reproduced with permission from Deakin University.

REFERENCES

Australian Collaborative Education Network (ACEN). (2015). *National strategy on work integrated learning in university education*. Retrieved from http://cdn1.acen.edu.au/wp-content/uploads/2015/03/National-WIL-Strategy-in-university-education-032015.pdf

Bowles, M. (2017). *Automation: Implications for skills packages*. Melbourne, Australia: DeakinCo.

Bowles, M., & Lanyon, S. (2016). *Demystifying credentials: Growing capabilities for the future*. Melbourne, Australia: DeakinCo.

Dede, C. (2017, December 11). Students must be prepared to reinvent themselves. *Education Week*. Retrieved from https://www.edweek.org/ew/articles/2017/12/13/students-must-be-prepared-to-reinvent-themselves.html

Deloitte Access Economics. (2017). *Soft skills for business success*. Melbourne, Australia: DeakinCo.

Ernst & Young (EY). (2012). *University of the future: A thousand year old industry on the cusp of profound change*. Retrieved from https://www.ey.com/au/en/newsroom/news-releases/australian-universities-on-the-cusp-of-profound-change---ernst-and-young-report

Manyika, J., Lund, S., Chui, M., Bughin, J., Woetzel, J., Batra, P., Ko, R., & Sanghvi, S. (2017). *Jobs lost, jobs gained: Workforce transitions in a time of automation*. New York, NY: McKinsey Global Institute.

Productivity Commission. (2017). *Shifting the dial: 5 year productivity review*. Canberra, Australia: Author.

World Economic Forum. (2016). *The future of jobs: Employment, skills and workforce strategy for the Fourth Industrial Revolution*. Geneva, Switzerland: Author.

Asheley Jones (ORCID: https://orcid.org/0000-0002-2441-6321)
Zena Consulting, Australia
Future Work Initiatives, Deakin University, Australia

JANICE ORRELL AND JULIE ASH

19. OTHERNESS IN PRACTICE (IN THE HEALTH PROFESSIONS)

EDUCATION FOR OTHERNESS AND CARING

Care [is] a necessary feature of being human ... (Heidegger, 1927/1962, p. 84)

This chapter elucidates the significance of an ethos and ethic of care in health professions practice, and the implications for a higher education curriculum for care in health professions education. We argue that the concept of "otherness" is central to engendering the practice of compassionate caring, and question how the university curriculum may encompass this disposition and its practices within the contemporary political, social and economic climate.

Commodification of people and services confronts tenets of traditional approaches to practice. It introduces new values that align more closely to business enterprises than service and care of the sick and vulnerable. Without care, the emerging affordances of artificial intelligence threaten to transform the "hands-on" personalisation of healthcare diagnosis and service provision. Increasingly, healthcare quality is measured for efficiencies, mediated by technologies, and governed by objective and detached laws and principles, thus robbing those being cared for of their personhood and individuality. The result? A risk of loss and devaluation of "wisdom of practice" derived from hands-on healthcare service delivery.

Education of the next generation of healthcare professionals is central to the cultivation and maintenance of a caring society. Yet, similar to the political and economic healthcare agenda, a demand for efficiency accountabilities and the uptake of emergent technological innovation (Dall'Alba, 2012) has also had an impact on higher education. An important challenge for education for caring in contemporary higher education is education for otherness due to the risk of notions of otherness and healthcare being lost in the current climate.

We have approached the generation of this chapter as a learning project. Our years of teaching in health professions courses in our university have been driven by our personal and strongly held core beliefs of social justice, mutual respect between teacher and learner, the importance of self-care in health and self-regulation in learning, and the centrality of compassionate care in healthcare practice. This chapter reflects merely a small part of the shared personal journey we have undertaken to arrive at new and deeper conceptual understandings of otherness and caring. This new appreciation of these complex concepts is followed by our critical reflection on the implications they hold for providing high quality healthcare practice and education for practice. Our first concern here is to unpack the meanings held regarding the key concepts of "otherness" and "caring", and how they are used in the justifications for contemporary health service practices. Following this, we argue for the importance of these notions within the contemporary climate of healthcare provision and the challenges that confound their explicit enactment in everyday healthcare practice. Finally, we explore a particular case for otherness and caring in educating health professionals to practise in Indigenous healthcare.

CONCEPTIONS OF OTHERNESS AND CARING

Notions of "the other" as a concept have various connotations across different disciplines. "Othering", in sociological terms, is used to describe a process inherent in identity development through social interaction. Those considered as "other" do not reflect our perceptions of who we are (Zevallos, 2011). Zevallos draws from the theoretical work of key sociologists (see Zevallos, 2011, for Mead, 1934 and Bauman, 1991) to describe othering as the social constructivist process of identifying "similarity" and "difference", through which we achieve a social identity and belongingness. It is also a function in the process of group formation in which members must meet certain criteria. Othering defines ourselves as much as others, denoting who does and does not belong. It is used to effect particularly by those already advantaged who possess social power and influence, and marginalises those who do not. In this process, those who are other lose their individuality; they are dehumanised, being defined and understood in terms of their membership of a group, not by their own unique attributes, capabilities, dispositions and complex life worlds. Differentiation through othering, while helpful in as much as it helps us form social hypotheses and predictions, is also argued to be the basis of social discrimination and marginalisation.

The interests of this chapter are not formed around this sociological notion of "othering", but rather they focus on "otherness". A notion of otherness rather than othering is used here to capture a capacity central to the human expression of empathy and compassion. An otherness disposition reflects a hermeneutic worldview in which the other is interpreted in terms of their complex life world. When the disposition of otherness is valued within practice, the intention is to understand what the life world means to the individual, the values they hold, the meanings they make and their responses to events in their lives (Clark, 2008).

Otherness is not an innate capacity in humans but can be acquired with maturity. It develops with emerging self-understanding, self-knowledge and self-identity. Otherness is both a disposition and a capacity that enables individuals to look beyond themselves and their own life experiences to imagine others' experiences, psychological states and personal responses. This capacity's emergence is fully understood in the act of "being", "becoming" and going beyond one's own ideas, feelings and values. It is observed in the acts of engaging with (Rydlo, 2010), and listening to, the other with empathy and compassionate caring (Koskinen & Lindstrom, 2013; Whitehead et al., 2014).

Attempts to define care distinguish between care and caring, with care being the values and concerns involved in caring, and caring being the ways and practices performed in providing care. Caring, like otherness, according to Gherardi and Rodeschini (2016), is not an innate human capacity. They describe caring enacted within caring professions as an organisational competence; a situated knowing embodied in everyday tasks. Care or caring in this sense is distinguished from the "care of" that defines services. Caring is emergent between people and things involved in care, shaped by the present situation but often tacit. This tacit emergent nature helps explain the difficulty theorists have in defining caring (Tronto, 2005) and the difficulty of anchoring its focus in curriculum, learning and instruction. Theorists describe constructs variously as the science of caring (Rydlo, 2010) or an ethic of care (Gherardi & Rodeschini, 2016), and distinguish between a duty of care and caring (Noddings, 1988, 2005). Whitehead et al. (2014) argue that caring is often construed as individual practice and something to be added, which risks its incorporation into the everydayness of practice. Concern about care is often conflated with concerns regarding power imbalances between the provider of care and the cared for. Ghreradi and Rodeschini (2016) describe care as:

> ... an emergent capacity of a cultural system ... which produces ethical conceptions of what is involved in care and what is not, which attitudes and behaviours towards care are appropriate and how people are involved in/by what they do 'as care'. (p. 268)

They explain that care is an organisational competence; a situated knowing that a group of professionals enact while attending to their everyday tasks, and that the practice of care evolves and emerges in the engagement of the carer and the cared for, and their mutual process of becoming related is "sustained by a collective competence that may be learnt and improved while practicing" (p. 268). Tronto (2005) agrees that care is difficult to define because of its situated nature and avoids defining it, instead describing caring as five phases of care essential for "the integrity of care" and used to judge the adequacy of care: caring-about, caring-for, care-giving, care-receiving and caring-with. Zembylas, Bozalck, and Shefer (2014) report the critical element of each phase as follows:

- Caring-about requires *Attentiveness*: involving recognition of others' needs in order to act.
- Caring-for requires *Responsibility*: involving responsible action in undefined ambiguous conditions as they arise in situations.
- Care-giving requires *Competence*: compassion and engagement in noticing and accepting responsibility for providing care is insufficient. Competent care-giving requires knowledge and skills that are appropriate and situated.
- Care-receiving requires *Responsiveness*: an understanding of, and concern for, the vulnerability of those needing care and the conditions of inequality through imagining being in a similar situation.
- Caring-with requires *Trust and solidarity*: moral qualities develop over time from the habits and patterns of care between carer and cared for.

Noddings (2005) describes care as basic in human life, and relational caring as "responding to each individual in such a way that we establish and maintain caring relations" (p. xv), which involves a capacity for appreciating otherness. Care is contextual; it is understood differently in the many individual and collective viewpoints that intersect in each instance of care. As a collective notion, care is important in guiding institutions, policy and curricula to create the context, affordances and culture for care and for learning care.

Thus far, we have considered otherness from an individual development perspective. Since caring occurs within the dynamic between self and others in the care context, we contend that understanding what it is to be caring in the way we enact healthcare and health education requires us to grapple with the concept of otherness both individually and collectively, such as occurs every day in healthcare wards or clinics, and in healthcare teams.

IMPORTANCE OF OTHERNESS AND CARING IN HEALTHCARE

The health professions are often referred to as the "caring professions". The intent may be caring, yet the systems that manage, if not control, the practice of health professionals are such that care delivery dominates caring practice. Health leaders and scholars argue that there is a systemic problem with what is deemed caring in the delivery of health services (Gaufberg & Hodges, 2016; Kelley et al., 2014; Neuwirth, 2002). These scholars point to 20th century rationalist projects to argue that the reductionist attempt to define, manage and systematise efficient and effective practice and its outcomes has favoured the measurable and tangible at the cost of the valued qualities of virtuous practice and caring. Neuwirth cites this as the paradox of modern healthcare: providing the best treatment and the worst care. Healthcare leaders may point to overarching mission and value statements as defining a context where the virtues of care are valued. The mismatch between espoused values and what is overtly valued in the managed care and measured practice context, however, leads to cynicism, contributes to practitioner burnout (Gaufberg & Hodges, 2016) and damages the trust of the public (Neuwirth, 2002).

Care matters, and there is evidence that caring, in terms of the quality of the physician-patient relationship, can improve healthcare outcomes (Kelley et al., 2014). Thus, care has demonstrable material as well as personal and social benefits, yet "one of the great challenges of modern medicine is to preserve the finest elements of caregiving in an environment that is increasingly dominated by market forces and routinized practices" (Kelley et al., 2014, p. 1). Health professionals see people at their most vulnerable when they most need care. How people respond to being vulnerable and in need of care is highly variable and unpredictable. Responses to care range from achieving proactive self-care to achieving a healthy practitioner-patient alliance to the depths of despair and passivity, or even anger and violence. The latter responses challenge notions of caring and stretch the limits of compassion and altruism. Caring is not necessarily easy or soft but asks us to engage with others in challenging situations where tensions and disparities abound, and our humanity is tested individually and collectively. Clark (2008) argues that in the human encounter with the other, especially in healthcare, the goal is understandings that are not unilateral reconstructions, but mediations shaped by each person's history and background. The mediations are replete with pre-understandings and fore-conceptions that must be negotiated for both to arrive at a new understanding. This is a complex process that relies on the ability to see what is questionable in the self, the other and in the imagination that fosters wisdom of practice while being open to new meaning (Clark, 2008).

According to Deloitte's 2016 survey of engaged healthcare (Betts et al., 2016), what matters most to consumers is that healthcare professionals understand them, know them and respect them, and that their healthcare is personalised. The Deloitte study found that healthcare consumers value the relationship between the carer and those cared for far more highly than either the cost of services or the availability of technological enhancement of care.

EDUCATIONAL RELEVANCE OF THEORETICAL PERSPECTIVES ON CARING

Gaufberg and Hodges (2016) argue for building humanism into health professional education to move beyond empathy and communication toward education that "involves learning to recognise and navigate tensions between values (empathy and objectivity, efficiency and quality, standardised and individualised care, for example) and to understand the ways in which power and privilege affect health care and learning interactions" (Gaufberg & Hodges, 2016, p. 264). Everyday care enacts caring competence, where professionals navigate the institution structures and affordances to individualise care; what good caring looks like is known collectively. However, we are less sure of how to best educate those entering the professions to be both competent and caring. Health professional education has not been immune from emphasis on rationalist solutions. There is concern that the outcome-based or competency-based curriculum movements have led to reductionist definitions of competencies, favouring explicit sets of knowledge and skills in which values and caring are subordinate, even where included or prefaced (Whitehead et al., 2014), and putting at risk integrated aspects of practice such as compassion and caring, and even wisdom (Gaufberg & Hodges, 2016). The dominance of psychometrics in assessment has reinforced the preference for measured outcomes (Hodges, 2013). However, as scholars worldwide embrace an assessment-for-learning programmatic paradigm, the discourse in assessment has shifted towards integrating qualitative, subjective expert judgements and providing rich feedback information to guide learner development (Schuwirth & van der Vleuten, 2011; van der Vleuten et al., 2012). This shift provides some basis for rethinking curriculum design, educational processes and assessed outcomes to embrace the concepts of otherness and caring.

The Case

So that this exploration of otherness does not remain forever locked in a theoretical void that is so far from actual practice that it cannot be transferred, we engaged in a conversation with Professor and Koori man Dennis McDermott, who occupies the Poche Chair for Indigenous Health and Well-Being at Flinders University's

Adelaide-based Poche Centre. Professor McDermott is also a National Senior Teaching Fellow, in which he has focused on eliciting individual and institutional transformation within Indigenous cultural safety education. A significant aspect of his work is the translation of his scholarship into education programs for university staff and students so they gain competence in practising compassionate caring that is culturally and clinically safe for Indigenous consumers of healthcare. In doing so, Professor McDermott recognises that when confronted with the realities and dilemmas of Indigenous experiences of "Othering" (Capitalisition [O] is his emphasis to denote difference), learners can experience disorientating complex emotions as their fundamental beliefs about identity, culture and "Others" are challenged. In response to this challenge, Professor McDermott speaks of the need for "de-Othering". By contrast, his use of the term "other" using the lower case "o" infers "difference requiring understanding", similar to the sociological concept of othering discussed earlier. The educational concept of de-Othering, through which to explore Othering and its consequences, is then seen as aligned with the concept of otherness discussed in this chapter.

A published poet, Professor McDermott knows the power of "story" and image in stimulating the emotional imagination needed to connect and understand, permitting engagement in critical reflection about identity, racism and colonisation. Indigenous poetry, visual images, art and narratives of othering are used liberally in the Flinders cultural safety program delivered to health students to provide palpable lived experiences rather than theory as vehicles for learning. Students reflect critically on taken-for-granted ways of thinking about white privilege, engaging in a transformative, discomforting and sometimes painful "unlearning" journey. The program has developed a successful assessment exercise that requires a scholarly level of critical reflection called *The Deconstruction Exercise*. Rather than answer the questions on Indigenous Australia offered up, anonymously, by previous student cohorts, students were directed instead to "identify assumptions, racialised language and/or approaches, and to identify omissions. A successful analysis will identify whiteness, institutional racism and an understanding of the social determinants of Indigenous health" (Sjoberg & McDermott, 2016, p. 30).

Professor McDermott was fully aware that the cultural safety program content focused on the historical and current lived reality of Indigenous Australians could be so confronting that it risks both student resistance and institutional disengagement. He deems the task essential, however, and he referred to Delany et al. (2016) to emphasise that becoming a culturally safe health practitioner requires an ongoing commitment to decolonise one's practice, along with the development of critical disposition. Professor McDermott's own scholarship has defined four ways students tend to respond to the Flinders cultural safety program: (1) Accepting and keen for more; (2) Moved, ashamed (nationally not personally), wanting to atone and uncertain; (3) Disturbed and flummoxed, resentful and feeling personally blamed; (4) Hostile/Rejecting, angry and disruptive of class (McDermott, 2017; Sjoberg & McDermott, 2016). He has found confirming, corresponding findings from Canadian First Nations scholarship arising from the *San'yas* Indigenous Cultural Safety programs.

Drawing on Indigenous traditions, Professor McDermott describes his response to students' diverse reactions as "holding" them while they transition to a point where they can courageously embrace the challenge of colonialism and its related othering of Indigenous people, and the inequality it has created. "Holding" (*kanyirninpa*), associated with cultural practices such as initiation, is a powerful educational concept arising from Indigenous knowledge about responsibly guiding a person through challenging and discomforting experiences in a manner that supports them to take courage, learn from wisdom and maintain connection.

When we considered Professor McDermott's practices of de-othering and otherness, we could see the elements of Tronto's framework of caring phases embedded in this education. We observed *caring about* the impact and value of the program for the learners, *attentiveness* to the needs of the other and the emotional pain of dissonance provoked in students, and *caring for* the students in Professor McDermott's exercise of *responsibility* to do something about the consequence of the challenging but necessary education. While this could be described as a "duty of care", which is largely legalistic and defensive, the processes applied reflected a deeper intent to ensure that the students would ultimately be able to function in culturally safe ways within undefined, unpredictable situations. A "competence" in the "care giving", grounded in pedagogical knowledge and skill that result from accumulated wisdom of practice which has been reflected upon transparently and critically, and supported by extensive scholarship, was also highly evident. There was also a *responsiveness* to the impact of the care provided ("care receiving"), evidenced in student feedback in which they articulated the pain they experienced (paraphrased by McDermott; after poet Tranströmer), "like walking through a wall" and "you can't go back through the wall" (McDermott, 2017, n.d.). In contrast to their expression of pain was knowing they were being "held" gently through a journey to uncover their white privilege and discovering the world is one. Finally, it was obvious that to achieve such outcomes in an institutional setting required acknowledgement that this was a shared journey ("caring with"), supported by the notion of "holding" related to Tronto's "trust and solidarity". The shared understanding is that de-othering and changing one's thinking habits are critical elements in providing healthcare and wellbeing for Indigenous Australians. Accomplishing this understanding and its place in everyday practice requires institutional engagement and support, plus individuals with the dispositions and competence to enact it in everyday practice.

Our conversation about otherness with Professor McDermott concluded with a shared acknowledgement that the features of this cultural safety program, of learning the need for de-othering and of holding learners safely while they confront the inequalities that result from their privilege, can be transferred to similar attempts to dismantle other forms of discrimination and inequality.

Challenges and Concerns

The idea of otherness is dangerous in the health professional education context because it challenges strongly held assumptions and raises emotions. Thus, we salute the bravery of our Indigenous colleagues in making no apologies for challenging students, engaging and "holding" them in transformative education that highlights the consequences of othering. That students are becoming professionals where value contradiction and false dichotomies pervade the hidden curriculum, is problematic. Knowledge and skill competence are far easier to make transparent than caring, with its emotional nature/consequences. Emotion is seen as clouding, not informing, rational thought. Emotional detachment is perceived as denoting "professional" care. Types of practitioners and patients become "others". Wallner (2010), however, argues that professionalism is aligned with virtue ethics and the "ethics of care", as a fundamental way of being involved with the other – a "form of being" – above knowing and doing. These ideals require system-wide strategies, and without them, conflict between healthcare practices and organisations' personal and social ideals can result in cynicism, empathy erosion, burnout and loss of public trust. An exclusively rationalist concept of professionalism risks practice becoming rule-driven, forgetting the humane aim of the rules.

DISCUSSION

The theories explored here provide fertile ground leading to an understanding that the challenge is to position otherness and caring as central concerns for educational institutions so that curricula transparently incorporate "otherness" and "care", not as a minor add-on, but integral to the complexity of education for care giving. Educational solutions include the explicit embedding of developing professionalism in assessment regimes and recognition of the importance of the psychosocial challenge for students of *becoming* professional. Curriculum implementation strategies include mindfulness of one's own and others' response to healthcare dilemmas and challenges, incorporation of critical reflective practice, and using narrative or drama to explore human and cultural dimensions of caring that intersect with the scientific and technological dimensions. Critical examination of the hidden curriculum is needed to make visible what is valued within the provision of health-related services.

Analysis of standards and curriculum documents from the College of Family Physicians Canada (Whitehead et al., 2014) found that statements depicting notions of compassionate care, including allusions and proxies, were largely absent from recent formal documentation. This absence reflects the impact of the organising curriculum framework of standardisation and accountability measures and competencies, despite assumed valuing of compassionate care. We question how can valued complex social constructs, such as caring, be included in standards and curricula? When translating curriculum standards or competency frameworks into *educational approaches,* we must not only attend to what is expressed, but also reflect on what is absent and what is hidden (Whitehead et al., 2014). If humanist, compassionate, ethical care is not expressed in curriculum documents, it can at least be expressed explicitly in educational processes. Caring must become part of the lived curriculum, promoting students' discourse of caring.

CONCLUSION

We argue that the concept of otherness enhances education for care through engaging the moral imagination in deep reflection of self, other, and self as other and enables an explicit inclusion of educational goals that aim to induce a deliberate focus on moral development which underpins caring professionalism. The concept of otherness provides a guide to critical reflection that is incorporated into educational activities to make what is largely absent or hidden explicit, thus empowers students' agency in becoming practitioners. Otherness is the antidote to the distancing associated with othering. Otherness embraces both identity and diversity. Imagining (or mirroring) the other's being provides the basis (respect, compassion and empathy) on which to engender competence in "care giving".

This pathway to becoming professional should begin at the start of students' educational programs. While efforts to educate for care necessarily begin within classrooms, students perceive a hidden curriculum from the earliest clinical exposure; a curriculum laden with messages about what it means to be a professional in the "real world". Thus begin the tensions students must navigate during their professionalisation. Individual teacher dispositions alone cannot achieve the ideals embedded in developing the capacity for caring. Institutional leaders' deliberate intentions to engender a culture of otherness and caring, as well as incorporating institutional practices that reflect those intentions, are essential.

ACKNOWLEDGEMENT

We sincerely thank Professor Dennis McDermott of the Poche Centre for Indigenous Health and Well-being for his generosity and wisdom.

REFERENCES

Betts, D., Read, L., Kaye, M., & Patton, A. (2016). *What matters* most *to the health care consumer.* Deloite Development LLC. Retrieved from https://www2.deloitte.com/content/dam/Deloitte/us/Documents/life-sciences-health-care/us-lshc-cx-survey-pov-provider-paper.pdf

Clark, J. (2008). Philosophy, understanding and the consultation: A fusion of horizons. *British Journal of General Practice, 58*(546), 58-60.

Dall'Alba, G. (2012). Re-imagining the university: Developing a capacity to care. In R. Barnettt (Ed.), *The future university: Idea and possibilities* (pp. 112-122). London, England: Routledge.

Delany, C., Ewen, S., Harms, L., Remedios, L., Nicholson, P., Andrews, S., Kosta, L., McCullough, M., Edmondson, W., Bandler, L., Reid, P., & Doughney, L. (2016). *Theory and practice: Indigenous health assessment at Australian Qualifications Level 9*. Sydney, Australia: Office for Learning and Teaching.

Gaufberg, E., & Hodges, B. (2016). Humanism, compassion and the call to caring. *Medical Education, 50*, 264-266.

Gherardi, S., & Rodeschini. G. (2016). Caring as a collective knowledgeable doing: About concern and being concerned. *Management Learning, 47*(3), 266-284.

Heidegger, M. (1927/1962). *Being and time* (J. Macquarrie & E. Robinson, Trans.). New York, NY: SCM Press.

Hodges, B. (2013). Assessment in the post-psychometric era: Learning to love the subjective and collective. *Medical Teacher, 35*, 564-568.

Kelley, J. M., Kraft-Todd, G., Schapira, L., Kossowsky, J., & Riess, H. (2014). The influence of the patient-clinician relationship on healthcare outcomes: A systematic review and meta-analysis of randomized controlled trials. *PLoS ONE, 9*(4), e94207.

Koskinen, C., & Lindstrom, U. (2013). Listening to the Otherness of the Other: Envisioning listening based on a hermeneutical reading of Levinas. *International Journal of Listening*, 27(3), 146-156.

McDermott, D. (2017). The uncomfortable road to cultural ease: Shifting focus to 'close the gap'. In G. Worby, T. Kennedy, & S. Tur (Eds.), *The long campaign: The Duguid memorial lectures 1994-2014* (pp. 126-135). Adelaide, Australia: Wakefield Press.

Neuwirth, Z. E. (2002). Reclaiming the lost meanings of medicine. *MJA, 176*, 21-79.

Noddings, N. (1988). An ethic of caring and its implications for instructional arrangements. *American Journal of Education, 96*(2), 215-230.

Noddings, N. (2005). *The challenge to care: An alternative approach to education* (2nd ed.). New York, NY: Teachers College Press.

Rydlo, C. (2010). Fighting for otherness: Student nurses loved experiences of growing in caring. *Örebro Studies in Care Sciences, 30*. Örebro, Sweden: Örebro University Publishers. Retrieved from https://www.diva-portal.org/smash/get/diva2:343237/FULLTEXT02.pdf

Schuwirth, L. W. T., & van der Vleuten, C. P. M. (2011). Programmatic assessment: From assessment of learning to assessment for learning. *Medical Teacher, 33*, 478-485.

Sjoberg, D., & McDermott, D. (2016). The deconstruction exercise: An assessment tool for enhancing critical thinking in cultural safety education. *International Journal of Critical Indigenous Studies, 9*(1), 28-48.

Tronto, J. (2005). An ethic of care. In A. E. Cudd & R. O. Andreasen (Eds.), *Feminist theory: A philosophical anthology* (pp. 251-263). Oxford, England & Malden, MA: Blackwell Publishing.

van der Vleuten. C. P. M., Schuwirth, L. W. T., Driessen, E. W., Dijkstra, J., Tigelaar, D., Baartman, J., & van Tartwijk, L. K. J. (2012) A model for programmatic assessment fit for purpose. *Medical Teacher, 34*, 205-214.

Wallner, J. (2010). How do we care for our future caregivers? Rethinking education in bioethics with regard to professionalism and institutions. *Medical Law, 29*(1), 21-36.

Whitehead, C., Kuper, A., Freeman, R., & Webster, F. (2014). Compassionate care? A critical discourse of accreditation standards. *Medical Education, 48*(6), 632-643.

Zevallos, Z. (2011). *What is otherness?* Retrieved from https://othersociologist.com/otherness-resources/

Zembylas, M., Bozalek, V., & Shefer, T. (2014). Tronto's notion of privileged irresponsibility and the reconceptualisation of care: Implications for critical pedagogies of emotion in higher education. *Gender and Education, 26*(3), 200-214.

Janice Orrell PhD (ORCID: https://orcid.org/0000-0003-1034-0642)
College of Education, Psychology and Social Work
Flinders University, Australia

Julie Ash PhD (ORCID: https://orcid.org/0000-0002-1569-6464)
College of Medicine and Public Health
Flinders University, Australia

THOMAS CAREY, FARHAD DASTUR AND IRYNA KARAUSH

20. WORKPLACE INNOVATIONS AND PRACTICE FUTURES

In both academic and workplace settings, there is a shared recognition of the need to prepare a workforce for future work environments. Stephen Lehmkuhle, founding Chancellor of the University of Minnesota-Rochester (an exemplar of innovation in undergraduate higher education) argued that work in the future will increasingly involve our graduates in working with knowledge, using knowledge practices and work in roles that all don't exist yet (in Zemsky, 2013). Similarly, from a workplace perspective, a leader in one of our local companies (Quan, 2018) has emphasised that while we don't know specifically what skills will be needed in future work, we do know that we'll need a high degree of adaptability and the capacity to navigate new territory and that ongoing worker development is essential. Amidst this uncertainty about future workplaces and work practices, the one thing we can count on is a growing focus on innovation and change in the workplace. Preparing learners for this future world in both higher education and workplace contexts will require us to focus explicitly on developing capability for workplace innovation which we regard as a meta capability (along with other cross-curricular graduate attributes) for adapting, shaping and leading future practices for the common good.

WORKPLACE INNOVATION AS A SOCIAL PROCESS

Workplace innovation goes beyond the notion of an innovation as a new product. It can involve changes in work practices across human, organisational and technological areas (Totterdill, Dhondt, & Boermans, 2016). In our institutions, our current workplace innovations focus on the social process of creating lasting value by mobilising new ideas in the workplace. We emphasise with our learners that innovation capability applies just as well beyond the workplace, in their other roles as community members and global citizens. The European Union Guide to Workplace Innovation notes that this social process "leads to significant and sustainable improvements both in organisational performance and in employee engagement and well-being" (Totterdill et al., 2016, p. 4). These dual goals for workplace innovation align with the theme *Practice Futures for the Common Good.*

ALL GRADUATES SHOULD ADD TO WORKPLACE INNOVATION

Our academic institutions are already engaged in developing innovation capability through programs in entrepreneurship and in social innovation leadership. However, these efforts engage few of our students. In our discussions with workplace innovation leaders – regional small and medium-sized enterprises and large national organisations in the corporate and public sectors – a recurring theme has been the need to engage every employee in contributing to innovation in the workplace. Table 20.1 illustrates one perspective on this goal. Adapted from workplace innovation research in Europe (Høyrup, 2012), the table shows how employees could take initiative for workplace innovation in progressively deeper ways. Initially, their role as contributors to innovation might be limited to adapting innovative practices developed elsewhere to accommodate their own local context. Over time, many employees will be able to take on roles with broader impacts, including contributing as Lead Users advising design teams on a pre-defined organisational challenge and initiating innovation projects by identifying and pursuing an innovation need. Many of our workplace partners also wanted to foster intrepreneurship for their employees.

Table 20.1. Levels of employee initiative in workplace innovation.

Adaptive Innovation	Employees engage in adaptation, development and diffusion of workplace innovations created elsewhere
Directed Innovation	Employees engage in designing and developing innovative solutions to identified issues and challenges, or in specific roles such as Sponsor Users
Proactive Innovation	Employees initiate and/or develop innovative solutions across a wide range of activities (Observe, Empathise, Ideate, Prototype, Test, Disseminate, etc.)
Intrepreneurial Innovation	Employees conceive, foster, launch and manage a new venture that is distinct from the organisation's current value proposition but leverages the parent's resources

This progression led us to view *Capability for Workplace Innovation* as a new graduate attribute for Practice Futures, a cross-curricular outcome which all of our graduates could be enabled to achieve. We provide examples of our recent pilot projects to highlight both the application of this approach and some of the surprising results we are applying in further work.

ENGAGING STUDENTS WITH TEACHING AND LEARNING INNOVATION

We want to prepare students for future practice where they can engage with innovation in many ways. One of the best routes to do this is to create *internal* experiential learning for innovation capability within our academic programs. Changes in our learning and teaching workplaces can become opportunities for students to develop their capability as knowledgeable critical friends of innovation in work practices, outcomes or products, processes, roles and expectations.

AN EXPERIENTIAL LEARNING OPPORTUNITY IN ADAPTIVE INNOVATION

We can illustrate this through an example experiential learning opportunity at the initial *Adaptive Innovation* initiative level in Table 20.1. Many of our instructors can easily cite examples where their introduction of a new learning practice has invoked a range of responses to innovation, from enthusiasm and support through scepticism and apprehension to reluctant compliance or even resistance (Baker & Hill, 2017; Ellis, 2015; Nguyen et al., 2017). Negative responses can be apparent in a classroom context, where students' "closed" body language shows their opposition.

HIST 2390 Pre-Class notes

Overview: When you think about future employment, you know employers will be interested in what you have learned in your studies. They'll want to know how what you learned can be translated into skills and knowledge you will put to work for them (and develop further).

Some of the skills and knowledge you will offer come from the specific subject matter of a course or major; some are developed across courses as essential employability capabilities that apply in any workplace, such as constructively critical thinking and effective team collaboration.

Recently some leading-edge organizations have begun to also look for skills and knowledge to adapt to – or develop – innovations in the workplace, involving new products, new work practices, new network structures for staff, clients, suppliers, etc. We expect there will be a growing need for this capability in the dynamic workplaces of the future.

A faculty team at our university is exploring how we can help you to develop that *capability to engage with innovation in the workplace* in our own 'workplace for learning'. At the same time, we're also working with employers in the region and across the country to keep in step with their changing needs. The attached graphic provides a map of our initiative.

Here's the question we want to explore with you in the class session next week:
How can I use my experiences "at work as a student" (in the work of learning) to demonstrate skills and knowledge that will transfer to other workplaces for future employers, with a special emphasis on the new capabilities emerging to address innovation in the workplace?

(We're also convinced that this capability to engage effectively with innovation will be equally applicable in your other roles beyond work, e.g., as community members and global citizens.)

Preparation:
- Skim the attached overview on *Building an Innovation-Enabled Workforce*. (to be discussed further in the class session)
- Watch this 7 minute video about the method we are planning to adapt to help you think about and demonstrate how your skills and knowledge will translate into the workplace. There's more about the SEAL process in the attached 3-page handout.
- Come prepared to discuss your experiences with innovations in your courses this term, such as the History Go! Assignment within HIST 2390

Notes on this Exercise:
- The in-class time included an introduction to Workplace Innovation and illustrations of the skills, knowledge and mindsets involved in Adaptive Innovation
- The "7-minute video", https://employability.uq.edu.au/plan-your-success/learn-your-experiences, is part of a free online course developed by the University of Queensland: EMPLOY101x "Unlocking your employability". The handouts for the SEAL process – Situation, Effect, Action and Learning – were also adaptations of University of Queensland employability resources [Richards & Reid, 2017]

Figure 20.1. Sample pre-class exercise and for experiential learning in Adaptive Innovation.

The introduction of a new learning practice can become a "teachable moment" for students to develop their skills, knowledge and mindsets in *Adaptive Innovation*. The pre-class handout in Figure 20.1 illustrates this process in preparing students for a History class by introducing innovation as a social process in the workplace – and in their other roles as community members and global citizens. The activity included reflection on their recent engagement in a History project to create content for a cell phone app which pops up location-aware anecdotes about interesting local historical places. The students had already shown their enthusiasm for the outcomes and process, although they had varied expectations when it became a capstone project.

EXPERIENTIAL LEARNING OPPORTUNITIES AND LEVELS OF INITIATIVE

Table 20.2 illustrates how we might translate the range of *Employee Initiatives for Workplace Innovation* from Table 20.1 into student activities within our teaching and learning workplaces. The table is an excerpt from a Scenario Map exploring how we could progressively introduce deeper levels of initiative and engagement in successive terms of an undergraduate program.

Table 20.2. Levels of initiative in innovation as student teaching and learning experiences.

Levels of "employee" initiative in innovation	Experiential learning opportunities within teaching and learning
Adaptive Innovation	Reflect on changes to personal practice from past work in learning
	Experience and reflect on more radical innovations in learning practice
Directed Innovation	Support innovations in teaching and learning practice in *Students as Partners* teams
	Support other learners in adapting Innovations in new learning practices
Proactive Innovation	Initiate, co-design and develop innovations in teaching and learning practice
Intrepreneurship	Initiate, design, gain support, develop and implement a new venture in teaching and learning

Many of the scenarios in Table 20.2 are extensions or re-framings of current practices in our teaching and learning. Here are some examples on which we are building:

- Introducing students to a more radical change in learning as an *Adaptive Innovation* activity might include a unit of study based on a short Massive Open Online Course (MOOC),[1] accompanied by reflection of both the student-as-user and student-as-adaptor experiences. This could include ways of adapting the MOOC for a wider range of users across a MOOC-Centred Learning Community.[2]

- One opportunity for *Directed Innovation* is explicit participation as Lead Users (Brem, Bilgram, & Gutstein, 2018) in innovation teaching and learning pilot projects. Senior students could support other students in their adaptation and reflection on new learning practices, like current Peer Mentor programs.[3]
- Co-development of instructional resources in Students as Partners programs (Mercer-Mapstone et al., 2017) is frequently an instance of *Directed Innovation* when a teacher identifies a challenge to be addressed with new and uncertain ideas. However, it is most valuable when students are engaged as full partners in all stages of the project for *Proactive Innovation* (Bovill, 2017).
- Other innovative teaching and learning practices engage students in deeper initiative for innovation through a mix of *Proactive Innovation* and *Intrepreneurship* activities. At Tufts University's Experimental College, students participate in designing, selecting and teaching our courses.[4] Every year Tufts students propose new courses on emerging topics to a joint student–faculty board, recruit instructors from the greater Boston area and participate as Peer Instructors. All students in the Bachelor of Science in Health Sciences program at the University of Minnesota – Rochester, design a capstone project (Neuhauser & Weber, 2011; Zemsky, 2013) for their final year which is a unique "set of learning experiences focused on an individualized theme that aligns with each student's personal and professional goals" (n.p.).[5]

In the current structure for these activities, students often do not receive guidance on how to translate these activities into the innovation context of an external workplace, or to integrate the skills, mindsets and contextual knowledge they have developed into a coherent framework of personal strengths and areas where they recognise the need for further development. Figure 20.1 illustrates how we might more explicitly reframe student learning experiences with innovation to address these needs.

PILOT PROJECTS WITH TEACHING AND LEARNING INNOVATION CAPABILITY

In engaging with workplace partners about their desired capabilities in workplace innovation for our graduates, one recurrent theme is *Learning to be Surprised*. Sometimes that is expressed as *Learning to Fail*, but careful planning of innovation pilot projects expects some surprises and does not consider a project to have failed if it provides evidence – positive or negative – about a key hypothesis underlying the planned innovation.[6] Three recent pilot projects to test initial ideas on developing student capability in workplace innovation are presented below.

REFLECTING ON ADAPTIVE INNOVATION

This pilot project in Humanities courses was introduced above within Figure 20.1. In addition to the History course, we repeated this in a course unit in first year English, including a focus on reading, writing and critical thinking skills in a university environment. The innovation in learning practices which students encountered in the English course was the use of ePortfolios to promote reflection and demonstrate cross-curricular skills (Light, Chen, & Ittelson, 2011).

What we learned: The students in both subject areas were intrigued by the idea of demonstrating and documenting a signature workplace capability which might help them to stand out in future interactions with employers. Almost all of the students had engaged with the online video case and in class were able to work through the SEAL analysis process steps we had adapted from the University of Queensland.

One obstacle for some students was their ability to recall their emotional response to the introduction of the innovation in learning practices, a key part of the Effects step in the SEAL analysis process. We had delayed students' reflection on the impact of the innovation to avoid competing for students' attention while they coped with the innovation. In retrospect this made it more difficult for some students who had not recognised their reaction as a critical incident for reflection. Recognising critical incidents is a key component of the process in learning from reflection on practice (Lindh & Thorgren, 2016). A final obstacle in fully engaging students in reflection on their Innovation experiences in these courses was the lack of an immediate, tangible mark of accomplishment. We offered to support students who wanted to incorporate their experiences with *Adaptive Innovation* in their ePortfolios. The Teaching Assistant leading this aspect of the project reported to us that our promise of future compensation in the form of Employability could not compete with more immediate student priorities. This contrasts with the student interest in digital badges in the case study description below as a tangible marker of achievement.

In the next iteration of our effort to promote reflection on *Adaptive Innovation*, we intend to implement changes which address these obstacles. These include a credit-bearing reflection option – a transition from secondary school to university learning program.

The reflection will be a credit-bearing option in another course unit at the same university, an Educational Studies course on the transition from secondary school learning practices to university level learning (which is a recommended course unit for all incoming Faculty of Arts students). This is part of a two-course sequence with the second course targeting senior students on the transition from university into the workplace. Another model we will explore is the recognition of employability achievement outcomes (via the Student Transformative Learning Record Model) at the University of Central Oklahoma (King, Kilbourne, & Walvoord, 2015), where students seeking to advance achievement on UCO's graduate attributes are able to identify course units to allow them to engage and develop their *Adaptive Innovation* capabilities.

INNOVATION TO SUPPORT DESIGN THINKING IN SOCIAL PURPOSE

This second pilot project focused on a new course in *Design Thinking and Innovation*. We wanted to see how existing pedagogies for Design Thinking might be extended to situate Design Thinking in a broader context of workplace innovation and to build an understanding of the multiple roles which employees might undertake in a Design Thinking project. The workplace context was an educational program affiliated with the host university. However, the setting contained several unfamiliar aspects for the students; the Tsawwassen Farm School[7] is a program for Sustainable Agriculture with a very applied focus based in an Indigenous community within a larger metropolitan urban area. Students could apply some of their understanding of educational institutions but were also forced to rethink some of their preconceptions and look at the situation with fresh eyes.

There are now multiple approaches for teaching Design Thinking to non-designers whose primary career interest is not in design, from early work (by Goldman et al., 2012) to more recent comparative studies (by Moseley, Wright, & Wrigley, 2018; Rekonen & Hassi, 2018; Wrigley & Straker, 2017). In that context, the base pedagogy approach for this course was close to that of García-Manilla et al. (2019) in combining Design Thinking as a structure for user-centred design with specific additional methods to foster creative thinking.

The Design Thinking process deliverables were focused on activities for *What is, What if, What wows* and *What works* (although due to time limitations the low-fidelity prototypes generated by the students to test *What works* were assessed for feedback by the Farm School managers rather than undergoing multiple prototype-test iterations). This was one of the constraining aspects which positioned these projects in our *Directed Innovation* category of Initiatives with prescribed roles and boundaries – as another example, the students identified challenges in collaboration with the Farm School managers but had limited access to Farm School students or customers for the retail operation.

As part of our larger agenda to pilot methods to develop capability for workplace innovation, we chose to also expose students to alternative ways to structure a Design Thinking process and to create awareness of the limitations of their knowledge about Design Thinking. We wanted to ensure the students appreciated that Design Thinking was more than a set of Skills for know-how; successful Design Thinking projects also require some measure of contextual knowledge about which Skills apply in a particular situation (know-why) and a Design Mindset (know yourself). The latter was in contrast to a Decision Mindset, identified by Rekonen and Hassi (2018) as the critical impediment experienced by novice Design Thinking teams. Building on the Employability aspects from the previous pilot projects, we provided students with an online case study from a corporate environment. IBM Enterprise Design Thinking[8] is a highly regarded exemplar of Design Thinking scaled across large enterprises. Deployment of Enterprise Design Thinking has delivered stellar improvements to the financial bottom line[9] along with the promise of ongoing organisational transformation.[10] IBM made available to the students the online learning module for a first level micro credential, the IBM Enterprise Design Thinking Practitioner digital badge. The students were asked to work through the model and analyse the similarities and differences between the corporate process and the one in which they had engaged at the Farm School, and assess their strengths and areas for improvement (via a mock job interview with an IBM hiring manager).

What we learned: The students responded very positively to exposure with a corporate Design Thinking environment. All students took the option of acquiring an Enterprise Design Thinking Practitioner micro credential and accompanying digital badge, indicating base level awareness of the Design Thinking process as implemented by IBM, as well as completing this Case Study assignment and discussing it in class. The students' results were mixed. Some students clearly picked up on the limitations of their Design Thinking experience and were realistic about how it could be extended in a supportive and knowledgeable working environment:

> *This class, it's just the beginning of design thinking. It's mainly helping every student to try to think about having empathy for the user. This IBM program is diving into every situation, exercises and successes. It's taking the next step and learning more about design thinking and listening to real situations.*

> *In this class we're encouraged to be creative in our own process and not be constrained by a 'right way of doing it'. In the IBM case study, we learned the benefits of some measure of standardization to allow teams to quickly establish working patterns and roles.*

> *Some improvements in my design thinking capabilities would be creating open ended questions for clients, encouraging them to tell a story.*[11]

In other cases, it was clear that the students had missed much of the emphasis on the finer points of team collaboration at scale, where the IBM case study was strong.

> *In my opinion, as long as a group of people like each other and get along well, they can accomplish many great things.*

> *Our textbook explained the steps of the design process by using very simple, catchy short stories – to which I could relate very well ... The IBM videos were interesting but very formal. I felt like explaining prototyping by telling a story of a child and his Lego block is just so much more inspiring.*

We'll certainly be using the case study again based on the positive reaction from students, while at the same time experimenting with other ways to maximise its potential value in exposing students to professional practice with Design Thinking at scale. One conclusion from our preliminary analysis is that the case study might have been more effective if it had been integrated into the course unit on a step-by-step basis rather than as a supplementary exercise at the end. Some students noted that much of the case study interactivity assumed that the participants were actively engaged in a team design project, and that some of the issues raised and suggestions offered in the case study module might have worked better during the project.

KNOWLEDGE MOBILISERS IN WORKPLACE INNOVATION

The previous project illustrates the logistical challenges in providing students with more comprehensive experiences of a workplace innovation project in an external context. In the early stages, students can bring fresh ideas and new perspectives on which employees can then build. However, ongoing student involvement in later stages of an innovation project should provide continuing benefit to the external workplace partner if we want to sustain effective partnerships over multiple iterations of a course unit. To test other ways that our students could provide benefit to innovation projects in external workplaces, we explored a pilot project with several potential workplace partners in our region (through the Innovation Committee of the local Chamber of Commerce). One of the authors was scheduled to lead a senior elective course unit on *Special Topics in Psychology* which could be tailored to specific areas. In this offering, the course subtitle was chosen as *Psychology, Design and Workplace Innovation*. Workplace Innovation as an area where students could work with external partners to mobilise research evidence in particular workplace contexts – for example, in particular work domains such as Construction and Healthcare or in a particular context such as a family-owned firm or other type of SME (Small and Medium-size Enterprise). This was intended to be a *Directed Innovation* role which would expose students to an innovation project they could bring distinctive value to by adding a capability to the existing team.

The potential SME workplace partners expressed strong interest in a pilot project along these lines. The research topics in the resulting course unit were structured to include both Workplace Innovation and Human-Centred Design – including Design Thinking – partly to link these topics in the students' minds and partly to provide a broader range of knowledge areas where students could bring value to external workplace partners. A key element of the student workload was a capstone project where the evidence from Psychology research could be applied to a setting external to the teaching and learning environment.

What we learned: To our delight, about half the students in the class chose to focus their capstone projects on Workplace Innovation. *All* of these students elected to develop their own projects in *Proactive Innovation* roles rather than pursue the default opportunity we had created for *Directed Innovation* – which we had assumed would be an easier way to engage with an external workplace.

PROACTIVE INNOVATION CATALYSTS IN STUDENTS' WORKPLACES

At the university hosting the *Special Topics in Psychology* course unit described above, most students commute from home and work part time. For many institutional purposes, this was regarded as a liability because it restricted on-campus activities while not often being career-related. However, the students who undertook a capstone Workplace Innovation project for *Special Topics in Psychology* found their part-time employment to be an advantage – providing a natural work-integrated learning site for them to engage with workplace innovation. All of them chose to identify and develop a workplace innovation project in their part-time work environments. For some students, the course experience prompted them to identify a need within their workplaces. Other students noted that the course gave them an opportunity to take initiative in proposing a Workplace Innovation which had already been on their minds but on which they held back, feeling constrained by their position as junior employees. The following capstone projects illustrate the variety and impact of these work-integrated learning experiences in workplace innovation:

- Redesigned the cleaning process and invented a new cleaning tray for camshafts at an auto parts corporation
- Designed and implemented a new training interface for customer service representatives at a major financial institution
- Designed and implemented a "bot" to answer frequently asked questions for the university-wide student association
- Designed and implemented an employee scheduling application for a major North American retail fashion company
- Prototyped a new online system that would allow front-line staff to communicate with managers at a local restaurant chain.

What (else) we learned: Because we had not foreseen the availability and appeal of these opportunities, we had not prepared for these new work-integrated learning contexts e.g. mentoring and supporting workplace proctors, and setting up complementary assessments from the academic and workplace perspectives. Another downside of this context shift was that each project was undertaken by an individual student rather than a student team as in the previous Design Thinking pilot. All of this will have to be addressed if we are to take full advantage of students' part-time work as sites where they can engage with workplace innovation and potentially take more initiative within this external-but-familiar environment.

TRANSFERRING CAPABILITY FROM ACADEMIC TO WORKPLACE CONTEXTS

Table 20.3 situates each of the pilot projects within the Initiative levels of Table 20.1 and the opportunities within Table 20.2 to engage with them in our teaching and learning environments. The final column of this table highlights how this could lead to work-integrated learning with innovation in external workplaces, including some of the possibilities occurring from the pilot projects. Our initial conception of the Initiative levels in Tables 20.1 and 20.2 was very linear, with a progression toward more initiative over time and a movement from innovation in the teaching and learning environment to innovation in external workplaces. The prominence of part-time external workplaces disrupts this view and enriches the space of possibilities, providing settings where students already have familiarity with the workplace context. We still believe there are benefits to students from engagement in a Directed Innovation role as catalysts for mobilising evidence from research in workplace innovation, especially in the context of a *Research Topics* course unit (despite the student-led pivot within the *Research Topics* course example toward more *Proactive Innovation*). Our interactions with potential workplace partners confirmed that there is significant interest in such roles, and that they could provide valuable career-related experience for students.

Table 20.3. Integrating curriculum units with experiential learning in workplace innovation.

Levels of Initiative in Innovation Projects	Experiential Learning Opportunities within Teaching and Learning	Sample Pilot Projects to develop Workplace Innovation Capability	Sample Experiential Learning in External Workplace Innovations
Adaptive Innovation	Reflect on changes to personal practice from past work in learning	Pilot project 1: ENGL 1100 and HIST 2490 reflection units	Reflect on innovations experienced in external part-time work practices
	Use and reflect on more radical innovations in learning practices		
Directed Innovation	Support innovations in teaching and learning practice in *Students as Partners* teams	Pilot project 2: DESN 2004 Design Thinking + external workplace case study	Collaborate with social purpose organisations in Design Thinking projects
		Pilot project 3 v1: PSYC 4900 knowledge syntheses with workplace partners	Collaborate in mobilising knowledge for external workplace innovation projects
	Support other learners in adapting innovative learning practices		Supporting adaptive innovation in learners' part-time workplaces
Proactive Innovation	Initiate and develop innovations in teaching and learning practice	Pilot project 3 v2: PSYC 4900 student-led innovation	Catalysts for innovation projects in learners' part-time workplaces
Intrepreneurship	Initiate, design, gain support and implement a new venture in teaching and learning		Collaborate with intrepreneurial projects in external workplaces

PRACTICE FUTURES AND INNOVATIVE WORKPLACE LEARNING

To train the next generation of … innovators, institutions of higher education must become labs for learning and engagement as they develop curricula that support an ecosystem of experiential learning opportunities that bridge gaps not only across disciplines, but also across social, political, and economic cleavages. By balancing theory and practice in pedagogy, educators can ensure a willing cadre of … innovators ready to tackle our most intractable problems. The future depends on how effectively we respond to the problems we face and to those we have not yet imagined. (Shah, 2016, p. 4)

Some concluding thoughts are in order regarding the implications of this kind of work within academic environments. The development of the pilot projects illustrated above began with scenarios of learning experiences and then looked for feasible ways to incorporate these experiences into curriculum planning. This "experience first" approach is instructionally well-aligned with a learning-by-doing approach for developing innovation capability. We are reminded, as Parker Palmer argued eloquently in his seminal book *The Courage to Teach*, that *how* we teach is a key part of *what* we teach (Palmer, 2007). We shouldn't expect to convince students of the importance of innovation in their own lives if they don't see it evident in our own, personally as instructors and organisationally as institutions. The emergence of micro credentials such as the STLR approach is an example of innovative instructors and an innovative institution coming into alignment to "practice what we preach". There are other innovations we need to track in new approaches such as Agile Learning Design (Arimoto, Barbosa, & Barroca, 2015) and Design Thinking for Learning Professionals.[12]

Our engagement with workplace partners at the leading-edge of workplace innovation is another area of innovation in our academic practices, where a new form of collaboration is emerging in which academic and workplace partners jointly work on the shared challenge of preparing a workforce for future environments where innovation, adaptability and resilience will be key requirements. We are exploring how we can develop instructional resources for workplace innovation which can be adapted by both academic and workplace partners, in much the same way that we have begun to use resources from IBM and other partners within our own classes. There are also other strands of academic development which we have not yet begun to leverage. As one example, the progress of initiative-taking in Table 20.1 does not yet incorporate results from research on Deliberate Professionalism and Agency as applied in the realm of Workplace Innovation. If we intend to provide graduates with the capability to shape and lead innovation in their professional practices – and not just adapt to innovations arising elsewhere – we must give our learners opportunities to exercise initiative and agency in our teaching and learning environments:

> A pedagogy of deliberateness will foster active participation in developing future practice, if students are given the opportunity to take ownership of decisions made in shaping the path, impact and nature of their own practice … and to explore and experience deliberate conduct by immersing themselves deeply and critically in professional knowledge practices … and community partnerships … even if it leads to experiencing uncomfortable learning and teaching. (McEwen & Trede, 2016, p. 227)

ACKNOWLEDGEMENTS

The work reported here reflects the collaborative efforts of a much larger team, including student research assistants (Anya Goldin, Natasha Lopes), the faculty members hosting the pilot projects mentioned (Jennifer Williams, Kyle Jackson, Iryna Karoush), our workplace partners (Blake Melnick, Karel Vredenberg) and the academic leaders who supported our work (Sal Ferreras, Diane Purvey, Carolyn Robertson and the British Columbia Association of Institutes and Universities). We also appreciate the examples from other institutions contributed by colleagues, including Monash University (Sarah MacDonald) and the University of Queensland (Deanne Gannaway, Dino Willox, Anna Richards).

NOTES

[1] E.g. *Innovation: The World's Greatest*, https://www.futurelearn.com/courses/the-worlds-greatest-innovations

[2] E.g. http://stemteachingcourse.org/mooc-centered-learning-communities-mclc/
[3] E.g. the Peers Ambassadors Leaders (PAL) program in the Faculty of Arts at Monash University, http://artsonline.monash.edu.au/transition/ambassador-program/
[4] http://www.excollege.tufts.edu/about Ex College offered its first courses in 1964, having been created by a Faculty Committee on Innovation and Experiment.
[5] https://r.umn.edu/academics-research/undergraduate-programs/bachelor-science-health-sciences/capstone
[6] The "Learning to be Surprised" phrasing comes from Jordan (2010); the hypothesis-driven design method for innovation projects is articulated in Horn, Crew, and Dribble (2018).
[7] http://www.kpu.ca/tfnfarm
[8] https://www.ibm.com/design/thinking/
[9] https://medium.com/design-ibm/a-new-study-on-design-thinking-is-great-news-for-designers-593f71b40627
[10] https://www.karelvredenburg.com/home/2016/8/30/design-thinking-whats-it-good-for
[11] Italicised quotes are data quotes which (being unpublished) do not have page numbers listed.
[12] https://www.bottomlineperformance.com/design-thinking-tools-learning-professionals-free-ebook/

REFERENCES

Arimoto, M. M., Barbosa, E. F., & Barroca, L. (2015). An agile learning design method for open educational resources. In *IEEE Frontiers in Education Conference Proceedings* (pp. 1897-1905). Washington, DC: IEEE Computer Society.

Baker, E., & Hill, S. (2017). Investigating student resistance and student perceptions of course quality and instructor performance in a flipped information systems classroom. *Information Systems Education Journal, 15*(6), 17.

Bovill, C. (2017) A framework to explore roles within student-staff partnerships in higher education: Which students are partners, when, and in what ways? *International Journal for Students as Partners, 1*(1), 1-5.

Brem, A., Bilgram, V., & Gutstein, A. (2018). Involving lead users in innovation: A structured summary of research on the Lead User Method. *International Journal of Innovation and Technology Management, 15*(03), 1850022.

Ellis, D. E. (2015). What discourages students from engaging with innovative instructional methods: Creating a barrier framework. *Innovative Higher Education, 40*(2), 111-125.

García-Manilla, H. D., Delgado-Maciel, J., Tlapa-Mendoza, D., Báez-López, Y. A., & Riverda-Cadavid, L. (2019). Integration of design thinking and TRIZ Theory to assist a user in the formulation of an innovation project. In G. Cortes-Robles, J. L. García-Alcaraz, & G. Alor-Hernández (Eds.), *Managing innovation in highly restrictive environments* (pp. 303-327). Cham, Switzerland: Springer.

Goldman, S., Carroll, M., Kabayadondo, Z., Britos Cavagnaro, L., Royalty, A., Roth, B., Hong Kwek, S., & Kim, J. (2012). Assessing d.learning: Capturing the journey of becoming a design thinker. In H. Plattner, C. Meinel, & L. Leifer (Eds.), *Design thinking research: Measuring performance in context* (pp. 13-33). Berlin, Heidelberg: Springer.

Horn, M., Crew, T., & Dibble, L. (2018). *Innovation management* (Entangled Solutions White Paper). Retrieved from https://michaelbhorn.com/2018/03/innovation-management/

Høyrup, S. (2012). Employee-driven innovation: A new phenomenon, concept and mode of innovation. In M. Bonnafous-Boucher, C. Hasse, & M. Lotz (Eds.), *Employee-driven innovation: A new approach* (pp. 3-33). London, England: Palgrave Macmillan.

Jordan, S. (2010). Learning to be surprised: How to foster reflective practice in a high-reliability context. *Management Learning, 41*(4), 391-413.

King, J., Kilbourne, C., & Walvoord, M. (2015). Student Transformative Learning Record (STLR): Capturing beyond-discipline learning in and out of the classroom. In *Research to Practice Conference in Early Childhood Education* (pp. 15-17). Edmond, OK: University of Central Oklahoma.

Light, T. P., Chen, H. L., & Ittelson, J. C. (2011). *Documenting learning with ePortfolios: A guide for college instructors*. Hoboken, NJ: John Wiley & Sons.

Lindh, I., & Thorgren, S. (2016). Critical event recognition: An extended view of reflective learning. *Management Learning, 47*(5), 525-542.

McEwen, C., & Trede, F. (2016). Educating deliberate professionals: Beyond reflective and deliberative practitioners. In F. Trede & C. McEwen (Eds.), *Educating the deliberate professional: Preparing for future practices* (pp. 223-229). Switzerland: Springer.

Mercer-Mapstone, L., Dvorakova, S. L., Matthews, K., Abbot, S., Cheng, B., Felten, P., Knorr, K., Marquis, E., Shammas, R., & Swaim, K. (2017). A systematic literature review of students as partners in higher education. *International Journal for Students As Partners, 1*(1).

Mosely, G., Wright, N., & Wrigley, C. (2018). Facilitating design thinking: A comparison of design expertise. *Thinking Skills and Creativity, 27*, 177-189.

Neuhauser, C., & Weber, K. (2011). The student success coach. *New Directions for Higher Education, 2011*(153), 43-52.

Nguyen, K., Husman, J., Borrego, M., Shekhar, P., Prince, M., Demonbrun, M., & Waters, C. (2017). Students' expectations, types of instruction, and instructor strategies predicting student response to active learning. *International Journal of Engineering Education, 33*(1), 2-18.

Palmer, P. (2007). *The courage to teach: Exploring the inner landscape of a teacher's life*. San Francisco, CA: Jossey-Bass.

Quan, G. (2018, August 19). We don't know what skills Canada will need in the future – and that's the point. *The Vancouver Province*. Retrieved from https://theprovince.com/opinion/op-ed/grace-quan-we-dont-know-what-skills-canada-will-need-in-the-future-and-thats-the-point/

Rekonen, S., & Hassi, L. (2018). Impediments for experimentation in novice design teams. *International Journal of Design Creativity and Innovation, 6*(3-4), 235-255.

Shah, S. (2016). Preparing public sector innovators through experiential learning. *Diversity and Democracy, 19*(3). Retrieved from https://www.aacu.org/diversitydemocracy/2016/summer/shah

Totterdill, P. Dhondt, S., & Boermans, S. (2016). *Your guide to workplace innovation*. European Union Workplace Innovation Network. Retrieved from https://ec.europa.eu/growth/industry/innovation/policy/workplace_en

Wrigley, C., & Straker, K. (2017). Design thinking pedagogy: The educational design ladder. *Innovations in Education and Teaching International, 54*(4), 374-385.

Zemsky, R. (2013). *Checklist for change: Making American higher education a sustainable enterprise*. New Brunswick, NJ: Rutgers University Press.

Thomas Carey PhD
British Columbia Association of Institutes and Universities, Canada

Farhad Dastur PhD
Department of Psychology
Kwantlen Polytechnic University, Canada

Iryna Karaush M. Advanced Studies in Arch.
Wilson School of Design
Kwantlen Polytechnic University, Canada

PART 4
REFLECTIONS

JOY HIGGS

21. REFLECTIONS ABOUT WORK

What Might Be My Future Practice Roles?

Why do we work? Barry Schwartz (2015) wrote about this question in one of the TED[1] (Technology, Entertainment and Design) international books that address *ideas worth sharing*. While for some people the answer could simply be to earn a living, he asks us to reflect on why some people find considerable job satisfaction beyond monetary gain while others are "actively disengaged" by their work.

> Satisfied workers are engaged by their work. They lose themselves in it. Not all the time, of course, but often enough for that to be salient to them. Satisfied workers are challenged by their work. It forces them to stretch themselves – to go outside their comfort zones. ... their workday offers them a measure of autonomy and discretion. ... [work is] an opportunity for social engagement. ... Finally, these people are satisfied with their work because they find what they do meaningful. Potentially their work makes a difference to the world. It makes other people's lives better. (ibid, p. 1)

This chapter invites readers to reflect on where you are in your work journey in terms of satisfaction, progression and contribution and where you would like to be, considering your own agency, external circumstances and possible work futures.

REFLECTION: DEALING WITH THE SHAPE THE WORLD IS IN

David Price (2013) has examined the global context in terms of:

> economic, social, environmental and political turbulence which has shaped our most recent past, and will dominate our immediate future. ... We face a complex set of possible futures and no one can authoritatively predict how things will look in ten years, let alone by the end of the century. We know only two things for certain. The first is that we should learn to embrace uncertainty, because this age of uncertainty could become permanent. The second is that if all the old certainties are gone, then we have to be open to radical shifts in how we work, live and learn. (p. 21)

Price (ibid) argues that these radical shifts are disrupting organisations, work, markets, societies and intellectual property strategies. He believes that this disruption needs to be dealt with by leaders, systems and people who want to face these challenges, in a way he describes as "going soft", that reflects generosity and support for the common good. It is fundamentally a movement of openness.

SOFT refers to four interconnected values (nouns) that are also actions (verbs). These are Share, Open, Free and Trust.

- **Share**. This relates to people engaging in common projects and discussion – often through the use of the Internet – via sharing of observations, skills, knowledge, insights and online expertise, such as information seeking. Sharing, it is argued, inevitably leads to collaboration. The power of online sharing lies in its speed and its capacity to support actions. According to Price, organisations are turning, increasingly, to collaboration to promote and progress innovation. Collaboration can extend beyond employees to "crowdsourcing".
- **Open**. This refers to values and actions that support the exchange of information, knowledge and skills. "Being open in the context of values appeals to people's sense of altruism and encourages reciprocity" (ibid, p. 55).
- **Free**. In this context the term free is multidimensional: it refers to cost-free, free to access and use, free to create, freedom from experts, free to roam (e.g. in learning), free to fail, free as a business model and free as entitlement. It follows from openness and leads to many new possibilities of engagement and learning.
- **Trust**. The first three values/actions rely on trust. Trust encompasses many aspects of work and engagement: the pursuit of honesty with clients and colleagues, trusting ourselves and other people, trusting governments, organisations and educational institutions to act honestly, trusting "in confidence" and privacy guarantees and trust in business.

While this model particularly refers to what people say or do online, it also reflects offline possibilities and momentum. Price (ibid) sees optimism in three main areas: the civic and political re-engagement of young people around the world, in the expansion of open transformations (e.g. scientific progress in agriculture, biotechnology, robotics) occurring across the globe, and in the way that the opening of learning is transforming many aspects of life. He also cautions against the misuse of the sharing, openness, free(dom) and trust which could lead to such negative outcomes as exploitation, unwanted effects of transparency, abuse of trust, etc. In adopting the strengths of any other model, there needs to be standards, safeguards and monitoring.

Reflection Point: *How might these ideas and strategies be used (or already are being used) in my work, my organisations and my footprint spaces?*

In his song *Saltwater*, Julian Lennon asks: What will I think of me the day that I die?[2] Rifkin (2011) poses the following challenge:

> Only when we begin to think as an extended global family, that not only includes our own species but all of our fellow travelers in the evolutionary sojourn on earth, will we be able to save our common biosphere community and renew the planet for future generations. (p. 270)

How do the ideas of (SOFT) Share, Open, Free and Trust provide guidance in dealing with other people in our work and the broader sphere of our lives? *What will*

WORK REFLECTION

we think of how we have dealt with others – not just those like us but more broadly – including people whom society marginalises, disregards or leaves behind? Consider, for instance, "others" and people who are marginalised in different settings. This topic is addressed by many authors and spokespeople. The English 1834 Poor Law Amendment Act[3] and more recently Katz (2013) in the USA differentiated between "the deserving poor" and the "undeserving poor". These titles prompted different levels of tolerance, support and justification of how people were treated and offered work or forced to work. We should pause to observe that marginalisation is a social pathology that reflects societies' inability to deal with diversity. Price's (2013) ideas about SOFT (approaches) as a mechanism to help society cope with future practice and employment uncertainties for diverse peoples, aligns well with a number of arguments in this book including the place of agency, social resilience and the way people's attributes and abilities influence their work futures.

SPACES AND TIMES OF PRACTICE

Reflecting on how each of us copes with future practice is an important challenge. Figure 21.1 provides a chronological interpretation of practice/work changes. Consider the changes you have already experienced on your way through your work, the types of changes you can imagine and potential futures.

	DOING	COMMUNICATING	KNOWING/THINKING	BEING	BECOMING
Traditional Early 20th Century Jobs	Practice-based	Traditional/limited communication	Growth of Propositional Knowledge	Jobs in primary/ secondary/ tertiary industries	Predictable
Established Late 20th Century Jobs	Evidence-based practice	Internet	Computerisation across most jobs	Blurring of "vocational" jobs	Rapid change
Re-invented Jobs in 21st Century	External scrutiny	Social media mobile devices	Shared decision making	Massive change in professionalism	Re-shaped, re-located
New Industry Jobs Early 21st Century	Wicked problems	Increase in media work and e-jobs markets	Partnerships, intermediaries	Supercomplexity	Fluidity, Flexibility
Emerging Jobs Early 21st Century	Entrepreneur-ship	Evolution of what counts as knowledge	Artificial Intelligence replacing/enhancing work/workers	The gig economy	Liquid, Peripatetic
Imagined Jobs 21st Century	Gender equity (improved)	Social media, Fake news Crowd sourcing	Epistemological fluency	Ontological re-invention	Evolved Boundary blurring
Unimaginable Jobs 21st Century		Creativity, Imagination, evolution, reflection, fire-catching, common good priorities, context instability, vulnerability, opportunities, diversity barriers diminished			

Figure 21.1. Work changes experienced, imagined and unimagined.

247

Reflection Point: *How do my ideas of possible, probable and preferable futures of practice and work relate to ideas above and to how my work and career is progressing? What role(s) might I have in addressing the common good?*

MY IDENTITY – IS THIS ME?

How does our situatedness shape us as individual with identity? Looking at our generations goes far beyond reflecting on our age and time in the workforce. Other factors – education, opportunity, cultural (and multicultural) aspects of society, workplace cultures, industries, etc. – impact on people's work choices, opportunities and experiences. Neugenbauer and Evans-Brain (2016) interpret the different generations and their work drivers (see Table 21.1). However, they warn that these labels, dates and interpretations are generalisations and advise us to review these ideas critically. Readers are encouraged to consider if these notions fit your background, work circumstances and aspirations and how you'd like to change.

Table 21.1. Generations and work.[4]

Generation	Career Drivers and Preferences
Veterans (those born between 1925 to 1942)	Orientation to work suggests they are loyal, work hard and prefer the status quo; … research suggests that Veterans build tacit knowledge (experience) relevant to their organisation (Eisner, 2005)
Baby Boomers (those born between 1945 and 1964, and entering the workforce from the mid-1969s)	Their values are said to be freedom from pressures to conform and seeking opportunities to learn new things (Jurkiewicz, 2000). As with the Veterans, Baby Boomers are more likely than their successors to have "bounded careers" (remaining in an occupational type or with the same employer) (Dries et al., 2008). They were more likely than other generations to find career-entry roles relatively easily, and didn't have clear career goals (Lyons et al., 2014).
Generation X (those joining the workforce in the early 1980s)	Their values are distrustful of authority and more loyal to their occupation/professions than to their employers (Johnson & Lopes, 2008). They are good communicators and problem solvers (Eisner, 2005). Careers are more likely to be mobile because of instability of employment opportunity (Dries et al., 2008).
Generation Y or Millennials (those born in the 1980s, entering the workforce in the late 1990s)	Their work values are said to be ambitious and they are eager to advance with earnings and status (Lyons et al., 2014). They are said to be more entrepreneurial than previous generations (Crumpacker & Crumpacker, 2007), needy, impatient (Johnson & Lopes, 2008), lacking focus and direction and low in problem solving skills (Smola & Sutton, 2002).
Generation Z or Post-Millennials or iGen (those born from the mid-1990s to the early 2010s)	This generation is staring to enter the workforce or are still at school. They have used the Internet from a young age, are typically comfortable with technology and engage in socialising through social media. Growing up through recent terrorist decades and the Great Recession (economic decline in world markets during the late 2000s and early 2010s) is considered to have given this generation a feeling of unsettlement and insecurity (Geck, 2007; Strauss & Howe, 1991).

The culture of industries and what is expected of workers and managers, varies across the primary industries (that deal with obtaining or providing natural raw materials and input to secondary industries, e.g. mining, agriculture or forestry), secondary industries that convert primary industry's raw materials into products and commodities for consumers (e.g. manufacturing) and tertiary industries (that are concerned with the provision of services). We are part of, and are shaped by, our work spaces and in turn, we can shape these spaces and even move beyond them.

Reflection Point: *How does the world see me as a worker or practitioner? How do I see myself? Am I shaped by my label or the box I'm placed in (e.g. my generation)?*

REFLECTION: MY PRACTICE AND DRIVERS FOR CHANGE

Chapter 1 introduced practice as an ontological and epistemological phenomenon, and social practice as comprising doing, knowing, being and becoming to pursue purposeful activities occurring within the social relationships of the practice context, practice discourse and the settings that comprise the practice world (Higgs, 1999). At this point we reflect back on this way of thinking about practice and on other ideas presented in this chapter using a force-field analysis

A force-field analysis is a device to reflect on planned change. The diagram below (Figure 21.2) is a hypothetical scenario. You can plan your own force-field change agenda using these ideas. To change, your aim is to strengthen the upward driver arrows and lessen, remove or negate the resisting downward arrows.

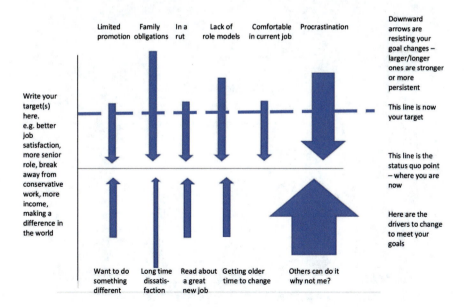

Figure 21.2. Using a force field analysis to plan changes to our lives and practice.

Reflection Point: *How satisfied am I with the way I am doing, knowing, being and becoming in my work and practice right now? Are there interests and goals I have that are driving me to change my practice? Where would I like to be in 5 or 10 years' time with work? How can I go about achieving these goals?*

REFLECTION: MOVING BEYOND THE SKILLS–JOB MATCH – WHAT MAKES WORK, WORK WELL?

In this final section a number of the themes across the book are drawn together in a figure that invites critical self-appraisal of our capability, choice, place, people, recognition, agency, timing and goals. The model below (Figure 21.3) proposes that these factors in various combinations can be arranged or pursued to address work concerns like: good job type matches for workers, desirable place and people matches for workers, other factors like recognition and agency that matter perhaps more to some people than others, and how timing of type of job or new job opportunities can be most pertinent considerations, or goals and aspirations might be primary drivers.

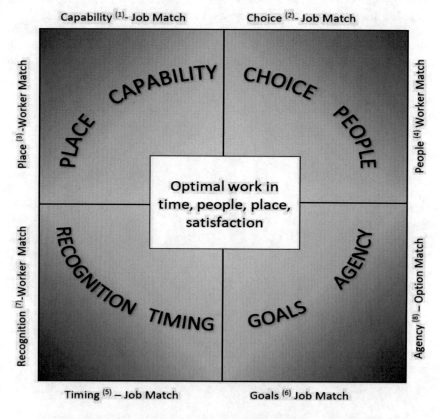

Figure 21.3. Making work, work well for me (and relevant others).

Figure 21.3 Notes
[1] Capability – includes qualifications, micro credentials, informal preparation
[2] Choice – includes job preference and choices in type of work/roles
[3] Place – includes location, work organisation, workplace
[4] People – includes employers/supervisors, co-workers, clients and their compatibility with the worker
[5] Timing – includes preparedness and preference for this job at this time and stage of career
[6] Goals – includes compatibility of work and organisation goals with the worker
[7] Recognition – includes respect shown to the worker, acknowledgement of their work contributions
[8] Agency – includes opportunities for and capacity of the worker to be agential and empowered

Reflection Point: *Where am I now in relation to my optimal work situation?*

PRACTICE FUTURES – WHO KNOWS?

This book has challenged hegemonic and status quo approaches and situations as well as complacency about future practices. It has presented ideas and hopes, uncertainties and concerns. Several authors have emphasised the danger of assuming the future will only consist of rearrangements of things we have seen and experienced in the past. They have encouraged us to "think the unthinkable" by considering futures that now seem improbable, or even unimaginable, but which are plausible if some currently emerging trends play out and interact in ways not seen before. By the same token, we must not overlook the likelihood that many of today's challenges and opportunities might still be around for decades to come, especially those relating to the psychological reasons why work has become such an important part of human life.

In all, we (the book editors) look forward to what the future of practice will be and hope that we, along with our authors and readers, will leave positive marks on these futures.

NOTES

1. https://www.ted.com/
2. Julian Lennon, *Saltwater*, 1991.
3. http://www.nationalarchives.gov.uk/education/resources/1834-poor-law/
4. The table created for this chapter is based on Neugebauer and Evans-Brain (2016) (rows 1–4) plus Geck (2007), and Strauss and Howe (1991).

REFERENCES

Crumpacker, M., & Crumpacker, J. D. (2007). Succession planning and generational differences: Should HR consider age-based values and attitudes a relevant factor or a passing fad? *Public Personnel Management, 36*(4), 349-369.

Dries, N., Pepermans, R., & De Kerpel, E. (2008). Exploring four generations beliefs about careers: Is 'satisfied' the new 'successful'? *Journal of Managerial Psychology, 23*(8), 907-928.

Eisner, S. P. (2005). Managing generation Y. *Advanced Management Journal, 70*, 4-12.

Geck, C. (2007). The Generation Z connection: Teaching information literacy to the newest Net Generation. In E. Rosenfeld & D. V. Loertscher (Eds.), *Toward a 21st-century school library media program* (pp. 235-241). Lanham, MD: Scarecrow Press.

Higgs, J. (1999, September). *Doing, knowing, being and becoming in professional practice*. Presented at the Master of Teaching Post Internship Conference, The University of Sydney, Australia.

Johnson, J. A., & Lopes, J. (2008). The intergenerational workforce revisited. *Organisation Development Journal, 26*(1), 31-36.
Jurkiewicz, C. L. (2000). Generation X and the public employee. *Public Personnel Management, 29,* 55-74.
Katz, M. B. (2013). *The undeserving poor: America's enduring confrontation with poverty* (2nd ed.). Oxford, England: Oxford University Press.
Lyons, S. T., Ng, E. S., & Schweitzer, L. (2014). Launching a career: Inter-generational differences in early career stages based on retrospective accounts. In E. Parry (Ed.), *Generational diversity at work* (pp. 149-163). Abingdon, England: Routledge.
Neugebauer, J., & Evans-Brain, J. (2016). *Employability: Making the most of your career development.* London, England: Sage.
Price, D. (2013). *Open: How well we'll work, live and learn in the future.* Great Britain: Crux.
Rifkin, J. (2011). *The Third Industrial Revolution: How lateral power is transforming energy, the economy, and the world.* New York, NY: St Martin's Griffin.
Schwartz, B. (2015). *Why we work.* London, England: TED Books, Simon and Schuster.
Smola, K. W., & Sutton, C. D. (2002). Generational differences: Revisiting generational work values for the new millennium. *Journal of Organizational Behavior, 23,* 363-382.
Strauss, W., & Howe, N. (1991). *Generations: A history of America's future: 1584–2069.* New York, NY: William Morrow.

Joy Higgs AM, PhD (ORCID: https://orcid.org/0000-0002-8545-1016)
Emeritus Professor, Charles Sturt University, Australia
Director, Education, Practice and Employability Network, Australia

NOTES ON CONTRIBUTORS

Note: ALTF refers to the Australian Learning and Teaching Fellows Network

Kristin Alford PhD
Director, MOD.
University of South Australia, Australia

Julie Ash PhD
Senior Lecturer in Clinical Teaching & Learning
Prideaux Research Centre for Medical Education
College of Medicine and Public Health
Flinders University, Australia

Ruth Bridgstock PhD, PFHEA, Member ALTF
Professor of Curriculum & Teaching Transformation
Centre for Learning Futures
Griffith University, Australia
Adjunct Professor, Creative Industries
Queensland University of Technology, Australia
www.graduateemployability2-0.com

Margot Cairnes MBA, BEd (Hons)
International Leadership Strategist

Thomas Carey PhD
Executive-in-Residence for Teaching and Learning Innovation
British Columbia Association of Institutes and Universities, Canada
Knowledge Management Institute of Canada
Principal Catalyst, Transforming Learning Together, Canada

James Cloutman MEd (SCU), PGCE, London Univ., BA (Hons.), King's College
PhD Candidate, Charles Sturt University, Australia
Member, Education, Practice and Employability Network, Australia

Megan Conway PhD
Chair, Health and Community Studies
Algonquin College in the Ottawa Valley, Canada

Steven Cork PhD
Crawford School of Public Policy
Australian National University, Australia
Principal, Ecoinsights, Australia
Director, Australia21, Australia

Farhad Dastur PhD
Department of Psychology, Faculty of Arts
2018 Teaching Fellow – Experiential Learning, Teaching and Learning Commons
Kwantlen Polytechnic University, Canada

NOTES ON CONTRIBUTORS

Peter Goodyear DPhil, FCIPD, Member ALTF
Professor of Education
University of Sydney, Australia

Joy Higgs AM, PhD, PFHEA, Member ALTF
Emeritus Professor, Charles Sturt University, Australia
Director, Education, Practice and Employability Network, Australia
https://www.practicefutures.com.au/

Debbie Horsfall PhD
Professor of Sociology
School of Social Sciences and Psychology
Western Sydney University, Australia

Graham Jenkins BA (Econ) (Hons.), FRCSA, MAICD, MLLR
Director, Graham Jenkins Pty Ltd

Asheley Jones DBA, Victoria University
Director, Zena Consulting, Australia
Senior Consultant, Future Work Initiatives, Deakin University, Australia
https://www.sfia-online.org/en/get-help/accredited-consultants/cvs/asheley-jones

Iryna Karaush B. Arch, M. Advanced Studies in Arch.
Faculty Member and Product Design Coordinator, Wilson School of Design
Coordinator, Design + Food Atelier
Kwantlen Polytechnic University, Canada

Rosemary Leonard PhD, BA (Hons.)
Professor and Chair in Social Capital and Sustainability
School of Social Sciences and Psychology
Western Sydney University, Australia
Coordinator of the Affiliation Group on Gender
International Society for Third Sector Research

Jennifer Malbon B. Interdisciplinary Studies (Sustainability) (Hons.)
Research Assistant
UNSW Canberra

Noel Maloney PhD
Lecturer, School of Humanities and Social Sciences
La Trobe University, Australia

NOTES ON CONTRIBUTORS

Lina Markauskaite PhD
Associate Professor in Learning Sciences
Co-Director, Centre for Research on Learning and Innovation
Faculty of Arts and Social Sciences
University of Sydney, Australia

Janice Orrell PhD, Member ALTF
Professor of Higher Education and Assessment
College of Education, Psychology and Social Work
Flinders University, Australia

Sandy O'Sullivan (Wiradjuri) PhD, Member ALTF
Deputy Head of School
School of Communication and Creative Industries
University of the Sunshine Coast, Australia

Daniel Radovich, Bachelor of Comms & Media
UX, UI & Front End Developer
https://www.danielrad.com.au/

Franziska Trede PhD
Associate Professor in Higher Education and Professional Practice
Institute for Interactive Media and Learning
University of Technology Sydney, Australia

Bernadine Van Gramberg PhD
Professor and Pro Vice-Chancellor (Graduate Research and Research Training)
Swinburne University of Technology, Australia

Paul Whybrow
Managing Director, Varda Creative Leadership
and Bodyboard Immersive Experiences
Creative Leadership Trainer for ADMA (Association for Data Driven Marketing and Advertising), AMI (Australian Marketing Institute) and MA (Marketing Association of New Zealand)
Professional Assessor for DeakinCo Microcredentials
www.vardacl.com